Building Feature Extraction with Machine Learning

Big geospatial datasets created by large infrastructure projects require massive computing resources to process. Feature extraction is a process used to reduce the initial set of raw data for manageable image processing, and machine learning (ML) is the science that supports it. This book focuses on feature extraction methods for optical geospatial data using ML. It is a practical guide for professionals and graduate students starting a career in information extraction. It explains spatial feature extraction in an easy-to-understand way and includes real case studies on how to collect height values for spatial features, how to develop 3D models in a map context, and others.

Features

- Provides the fundamentals of feature extraction methods and applications along with the fundamentals of machine learning
- Discusses in detail the application of machine learning techniques in geospatial building feature extraction
- Explains the methods for estimating object height from optical satellite remote sensing images using Python
- Includes case studies that demonstrate the use of machine learning models for building footprint extraction and photogrammetric methods for height assessment
- Highlights the potential of machine learning and geospatial technology for future project developments

This book will be of interest to professionals, researchers, and graduate students in geoscience and earth observation, machine learning and data science, civil engineers, and urban planners.

Building Feature Extraction with Machine Learning

Geospatial Applications

Bharath H. Aithal and Prakash P.S.

CRC Press is an imprint of the
Taylor & Francis Group, an **informa** business

First edition published 2023
by CRC Press
6000 Broken Sound Parkway NW, Suite 300, Boca Raton, FL 33487-2742

and by CRC Press
4 Park Square, Milton Park, Abingdon, Oxon, OX14 4RN

CRC Press is an imprint of Taylor & Francis Group, LLC

ISBN: 978-1-032-25533-0 (hbk)
ISBN: 978-1-032-26383-0 (pbk)
ISBN: 978-1-003-28804-6 (ebk)

DOI: 10.1201/9781003288046

Typeset in Times
by MPS Limited, Dehradun

Dedication

To all geospatial application enthusiasts out there

Contents

Preface

According to a United Nations report, global urban populations reached 55% in 2016 and are expected to reach 68% by 2050. However, due to the rapid increase in the rate of migration from agrarian to urban societies, actual populations will most likely be higher. According to the 2011 census, India has over 4,000 cities with populations of more than 100,000 people. The United Nations advocated for Sustainable Development Goals in 2015, which include sustainable infrastructure development as a primary factor. The United Nations has proposed 17 core themes for sustainable development that must be achieved by 2030. In order to ensure transparency and scale, Article 76 of the 2030 Agenda encourages the use of earth observation and geospatial data to aid and track the development of various projects. Building footprints, which include polygonal outlines of building roofs and height components, are included in the base datasets in urban settings. The standard aerial image, for example, covers 2.25 km^2 and contains 700 buildings in a low-density area. According to tax records, a city like Bangalore has a 740 km^2 administrative area with 25 lakh buildings. The actual figures have to be much higher. Updating these large datasets or creating maps is a difficult task. These factors inspired the authors to write this book, which explains state-of-the-art spatial data creation methods. The book also shows how to use geospatial technology and machine learning to assess rooftop solar photovoltaic potential, as well as a few other case studies.

The following are possible key takeaways from the book: deep learning approaches are the most popular choices for automated feature extraction. Stereo satellite images can provide building heights over a large geographic area. The most practical and appealing method of digitally portraying urban scenarios is 3D mapping. Geospatial technologies can be used to create large-scale social applications with minimal resources.

Despite the benefits of geospatial technology combined with machine learning, these technologies have some limitations. The first is the algorithmic constraint, which means that the models described in the book are suitable for images from similar sensors. If they are applied to images from unknown sensors, the results may be unsatisfactory. When applied to different geographic locations, the models tend to perform poorly; however, the model can be fine-tuned using a small number of training sets and the transfer learning method. High computation capacity, such as GPU, is required for model training. The spatial datasets are either difficult to obtain or expensive to purchase. Even if sufficient financial resources are obtained, there is a space policy barrier among several nations.

Machine learning models must first be generalized so that they can be applied to a variety of urban scenarios. Benchmark models need to be created so that a specific spatial feature can be extracted with minimal effort. To reduce manual post-processing, polygon and line features must be regularized using automated procedures. Terrain model or bare earth model preparation is currently a difficult task. Existing methods are ineffective for large areas. As a result, a terrain-dependent approach to digital terrain model preparation is required. For a fully automated workflow, the models must be tested for very high-resolution satellite images over a large geographical area.

This book aims to provide an overview of a structured workflow for producing critical spatial datasets using automated procedures.

Acknowledgements

Writing this book was both more difficult and more rewarding than we had anticipated. Without our respective families and friends, none of this would have been possible. It gives us great pleasure to acknowledge the help of others in completing this book. Several people assisted directly with the book's conceptualization, evaluation, and feedback, while others provided indirect support throughout the journey. Dr. Vinay, Dr. Chandan MC, Dr. Nimish Gupta, Aishwarya Narendr, Madhumita Dey, Manoj BS, Arpita, and several others in the Energy and Urban Research Group at the Indian Institute of Technology Kharagpur in the Ranbir and Chitra Gupta School of Infrastructure Design and Management deserve our heartfelt thanks for their support, knowledge sharing, and keeping the energy level high. Soumya Kanta Das, Mansi Unniyal, Shivam Bhosale, Sakshi Tyagi, Shafia Ahmad, Gaurav S, Nishant Rajput, Vishal S, Mohit Verma, Satarupa Mitra, Anshul, Jahnavi Soni, and many other research students who were part of the lab deserve special recognition for bringing innovative perspectives to the learning process.

We appreciate the review team's prompt suggestions, as well as the Ranbir and Chitra Gupta School of Infrastructure Design and Management and the Indian Institute of Technology Kharagpur's continuous support. A million thanks to Prof. Dilip Kumar Baidya, Prof. Tarak Nath Mazumder, Prof. Arkopal Kishore Goswami, Prof. Ankhi Banerjee, Prof. Bhargab Maitra, Prof. Joy Sen, Prof. Mukund Dev Behera, and Prof. A.N.V. Satyanarayana of the Indian Institute of Technology Kharagpur. We express our sincere thanks to the Energy and Wetlands Research Group at the Indian Institute of Science Bangalore for all the help and support. We thank Prof. Ramachandra T.V. for his unconditional support throughout our knowledge acquisition and dissemination journey. We would also like to express our gratitude to the IITKGP USA Foundation for their support in this journey.

Finally, we want to thank our families for their unconditional support in all of our endeavours.

Bharath H. Aithal and Prakash P.S.

Author Biographies

Dr. Bharath H. Aithal is currently an Assistant Professor in the Ranbir and Chitra Gupta School of Infrastructure Design and Management at the Indian Institute of Technology Kharagpur. He obtained his PhD from the Indian Institute of Science. His areas of interest are spatial pattern analysis, urban growth modelling, natural disasters, geoinformatics, landscape modelling, urban planning, open-source GIS, and digital image processing. More details on his publications can be found at:

Google Scholar: https://scholar.google.co.in/citations?user=j-trcFUAAAAJ&hl=en

Research Gate: https://www.researchgate.net/profile/Dr-Bharath-Aithal

Dr. Prakash P.S. is a postdoctoral researcher at the Irish Centre of High-End Computing, Galway, Ireland, working on geospatial technologies. Prakash has substantial experience with earth observation datasets, including remote sensing, drone-based imagery, surveying, spatial libraries, machine learning, and artificial intelligence technologies. He has worked in geospatial technology and the renewable energy industry for over four years. His other qualifications include a master's in technology in geoinformatics from Bangalore's Karnataka State Remote Sensing Application Center and a bachelor's degree in civil engineering from Bangalore's Rashtreeya Vidyalaya College of Engineering.

1 Introduction

1.1 GEOSPATIAL TECHNOLOGIES

The convergence of artificial intelligence, high-end computing with geospatial technologies is inevitable in the near future for the majority of areas. A large portion of geospatial technology is about data generation and information extraction. The core pillars of the geospatial domain, such as remote sensing, geographic information systems, global positioning systems, and intelligent solutions, centre around data gathering, information extraction, and decision-making. In this context, studies, extensive research, and innovations are needed in the geospatial domain to support future technologies such as autonomous vehicles, intelligent transport, cargo drones, smart cities, and much more.

Geospatial technology is an emerging field of study that includes geographic information systems, remote sensing, and intelligent sensors. However, we can't imagine any technology in the future without associating it with artificial intelligence and high-end computing. The domain has already witnessed significant automation penetration in satellite-based remote sensing, be it cloud detection or super-resolution methods (Schmitt et al., 2019). Prominent software tools, both open source and proprietary, extract features of interest using machine learning methods. Looking at this progress across the spectrum of geospatial technologies, it is inevitable to understand the overall process of spatial feature extraction from a technological point of view. When we say *spatial feature extraction*, this can mean many things such as the preparation of valuable datasets including building footprints, trees, road centrelines, water bodies, and many others.

We can understand the importance of geospatial data generation with one tragic past event. In 2017, Hurricane Maria in the northeastern Caribbean washed away nearly one million houses. After the hurricane, 5,300 volunteers spent 70 days creating data for the first response group involving digitization of about 9,50,000 or 950,000 damaged buildings, roofs, 30,000-km of roads, and several other physical infrastructures. Imagine a satellite image, an automated feature extractor tool that can accomplish this task on a high-end computing machine in a fraction of the time, which normally takes months to do manually by personnel. This involves using remotely sensed satellite or aerial images, feature extraction, and spatial data preparation within a geographic information system.

DOI: 10.1201/9781003288046-1

We need to understand this process in terms of technology and as a solution enabler. It involves the amalgamation of multiple technologies and requires expertise from different domains. A holistic approach is essential in understanding a larger picture. Spatial data in its primary form consists of three-dimensional geometry and extracting them are crucial to realize or develop real-world solutions. This book will look into aspects such as the source of geospatial big data, feature extraction using automated methods, object height estimation, 3D feature mapping, and connected applications. The book provideds overview of complete steps from data acquisition to applications by including relevant case studies.

1.2 FEATURE EXTRACTION

The spatial datasets in urban environment can come from various sources such as demographics, infrastructures, and many more. Land features such as roads, buildings, trees, and utility networks are well represented on digital maps. The current methods of map-based dataset generation happen manually. These may include physical data capture, digitization of remote sensing images, conversion of paper maps to digital formats, etc. Table 1.1 shows various techniques and data sources adopted in preparing map-based datasets. These sources and methods accelerate the data preparation process significantly; however, the next step is to automate mundane or manual processes existing currently. Research by Breunig et al. (2020) point towards future technological developments related to smart cities based on concepts of digital twins, web platforms, and big geospatial data.

TABLE 1.1
Preparing Map-Based Datasets: Techniques and Data Sources

Data Sources	Techniques	Data Outputs
Manual survey • Total station • GPS or mobile survey Satellite imagery or aerial imagery • Panchromatic images • Multispectral images • Hyperspectral bands • Microware data	• Manual digitization • Data conversion • 3D scanning • Photogrammetry • Image classification	• Atlas • Digital maps • Web GIS • Geo-analytics • 3D City models

1.3 GEOSPATIAL MACHINE LEARNING

Algorithm-based machine learning techniques in urban feature extraction are a highly focused area of research with numerous advantages over earlier methods in terms of speed, replicability, and model-based approach. Remotely sensed images acquired by satellites or drones typically deal with vast information. As a result, machine learning techniques are best suited for dealing with enormous geographic areas such as cities. Studies about large-scale land-use land cover mapping, land surface temperature estimations (Jeganathan et al., 2011; Bharath et al., 2019), and many other application areas have benefited by adopting automated machine learning techniques. However, the quality of results is questionable while dealing with the fine spatial features. Continuously growing volumes of data, quicker response time, increasing cost of the human workforce, and robust computational capacity have made machine learning prevalent in geospatial technology.

Machine learning-based classifiers such as maximum likelihood classifiers, random forests, support vector machines, and multi-layer neural networks can extract built-up areas efficiently. Still, they have limitations for extracting the proper shape of buildings from remote sensing images. Also, several artefacts and noise in the prediction are generated using automated methods. Table 1.2 lists the prominent machine learning techniques adopted in retrieving datasets from remote sensing imageries. However, machine learning techniques are adopted for several purposes: dimensionality reduction, sampling, resolution merge, object identification, and many others.

Deep learning concepts can be traced back to well before the widespread use of computers (Alom et al., 2018). However, the field has witnessed significant development with high-computing capabilities such as graphical processing units or GPUs. Deep learning is the subset

TABLE 1.2
Popular Geospatial Data Generation Machine Learning Algorithms

Machine Learning Techniques	Primary Uses in Geospatial Applications
Maximum likelihood classifier	Image segmentation, land-use land cover maps
Random forest	Land-use land cover, feature extraction
Support vector machine	Feature extraction, image classifications
Multi-layer perceptron	Image segmentation, feature identification
Deep neural networks	Classifications, feature extraction
Convolutional neural networks	Instant segmentation, feature extraction
Generative adversarial networks	Image generation, classification

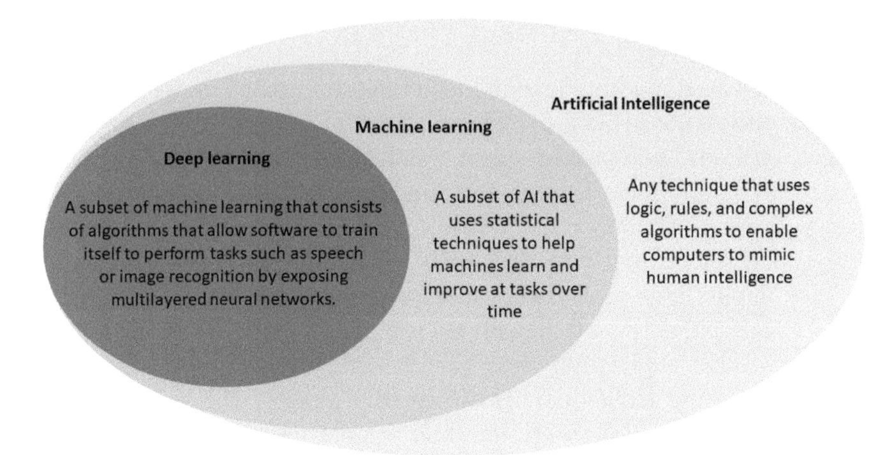

FIGURE 1.1 Concepts of deep learning, machine learning, and artificial intelligence.

of the broader area of machine learning and artificial intelligence, as shown in Figure 1.1. The research company ImageNet, in its third edition in 2012, saw a breakthrough in achieving an additional 10.8% accuracy points using deep learning concepts. The deep learning concepts attracted enormous attention thereafter for computer vision, speech recognition, and natural language processing. The studies of remote sensing imageries towards automated techniques have been increasingly popular ever since.

Object detection and segmentation are the two primary deep learning approaches used on remote sensing images. Object detection helps identify a specific object in an images, whereas segmentation helps classify every pixel in an image into a certain number of classes. Instant segmentation, a combination of segmentation and object detection, has been included as a new category of machine learning. Object detection can be used to identify and count spatial features such as trees, buildings, vehicles, and others things from remote sensing images. This technique produces a bounding box around an object with a certain confidence level. The object detection technique does not help generate the dimensions or area occupied by the features.

The segmentation technique classifies the entire scene, such as built-up roads, vegetation, water bodies, vehicles, etc. In addition, this technique helps to produce land-use and land cover classes using remote sensing imagery (Syrris et al., 2019). The feature extraction process from remote sensing images can be attributed to these two methods, generating vector-based spatial features with a definite area or length. To perform feature extraction, segmentation of images can be performed using binary classes.

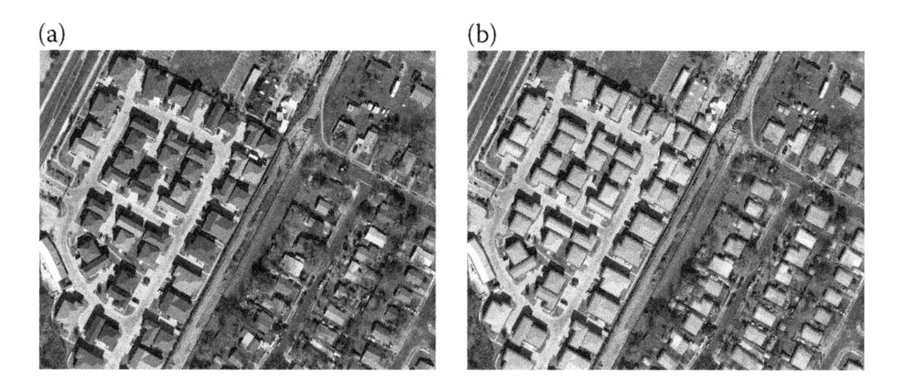

FIGURE 1.2 (a) Original image. (b) Image segmentation using remote sensing is depicted using buildings as an example.

Thereafter, segmented outputs are converted into vectorized features. The technical challenge is to achieve higher segmentation accuracy so that the polygonal outputs are consistent with real-world scenarios. The illustration of a segmentation output is shown in Figure 1.2.

1.4 HEIGHT ESTIMATION

The comprehensive understanding of urban structure comprises prominent land-use patterns, built-up details, building height, etc. Over the past several decades, urban development has been accompanied major buildings that are multi-storied or high-rise structures. This may be due to several reasons: the high cost of land, optimization of the construction cost of buildings, the type of occupancy, increase in plinth and plot area ratio, and several others. Since many buildings that constitute urban development are multi-storied, it is vital to consider the third dimension in quantifying urban structures. A bunch of high-rise buildings replacing a colony of old buildings can be witnessed across multiple cities. The redevelopment of slums is one typical example of a similar scenario. City developmental authorities are opting for multi-storied housing apartments and residential layouts because future construction needs to consider space optimization and vertical development.

Capturing the vertical growth of urban areas has become inevitable to evaluate the overall analysis of urban structures. Measuring vertical urban development could play a crucial role in urban planning, disaster management, environmental and ecological studies, energy assessments, communications, resource allocation, etc. In addition, urban built-up volume acts as a proxy measure for factors such as higher economic activities. Such data generation at the city scale is tedious, time-consuming, and labour intensive by traditional means. This makes it practically nonviable to gather urban structure data periodically, which is crucial for effective decision-making. The aerial surveys that capture overlapping photographs with films have helped to map 3D urban forms. Later, digital photogrammetric techniques proved efficient in producing needed datasets. The advent of satellite stereoscopy has made significant improvements over the past two decades in 3D mapping capabilities. Even though drones can produce precise outputs, the major bottlenecks are difficulties or restrictions involved with flying operations and expenses. Satellite stereoscopic methods are advantageous in cost, computational efficiency, and temporal capabilities to produce 3D maps of urban structures at various scales. However, the quality of the output depends mainly on the sensor's resolution.

Traditional maps containing earth features on a two-dimensional surface do not represent the 3D urban scenario's complete nature. To effectively capture and represent urban setup as it exists in the real world, 3D spatial datasets and sophisticated software tools are necessary. Among the prevailing methods are stereo satellite imagery, aerial images, and terrestrial data capturing methods. Remote sensing imaging means capturing information using regions of the electromagnetic spectrum such as visible, infrared, microwave, and laser beams. This book focuses on the visible region of remote sensing using high-resolution imaging sensors.

Based on the visualization and analysis needed, 3D city models are classified into levels of detail (LOD). The Open Geospatial Consortium (OGC) has organized the level of details of 3D models into five classes (Gröger et al., 2006). As per the definition, LOD1 contains the buildings' well-defined height with flat roofs. This chapter explains the background, methods, and intricacies related to the preparation of LOD1 with a focus on the vertical components of the buildings.

1.5 THREE-DIMENSIONAL MAPPING

The development of 3D maps using remote sensing technology has revolutionized how spatial data are conceptualized, collected, and analysed. The capability of geospatial methods to capture buildings, trees, towers, bridges, and many other structures within a 3D environment has provided much-needed datasets in a suitable format for several developmental activities. Significant research efforts have gone into remote sensing data collection such as platforms (satellite, aerial, terrestrial) or sensors (multispectral Lidar, radar, etc.). Also, technologies are rapidly evolving to automatically convert raw data into meaningful 3D maps. The megacities of developing countries extend in the third dimension and horizontal sprawl. The significant addition of built-up areas, construction of high-rise buildings, roads, and other infrastructures are the major

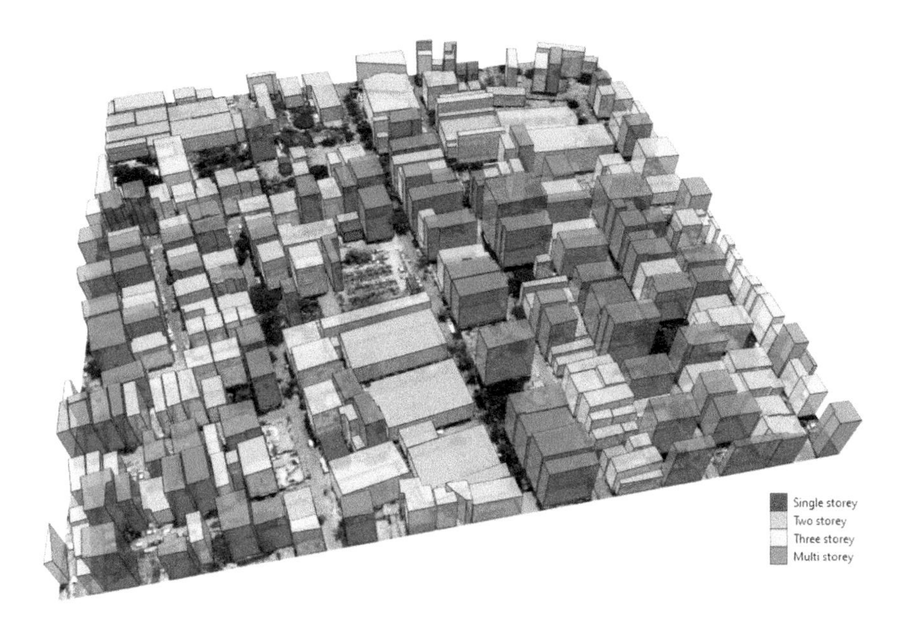

FIGURE 1.3 Three-dimensional model representation.

contributors to urban growth. Understanding and modelling these phenomena of urban development using 3D geospatial technology is helpful in sustainable planning and providing intelligent solutions.

The representation of a building on a map as a 3D object has always attracted urban administrators' interest. The third dimension within digital mapping simplifies feature representation and relationships among other spatial objects, as depicted in Figure 1.3. It is an intricate task to incorporate 3D spatial data models with other map features using topological relationships in a digital map environment. The urban area is complex and comprises many relatively small spatial objects such as buildings, parks, roads, streetlights, etc. The use of spatial technology to represent urban systems is dominant due to topological relationships. Compared to natural features such as trees, rivers, hills, etc., man-made objects are regular in structure, shape, and distribution. This characteristic assists developers in representing urban features as spatial objects with specific geometry.

REFERENCES

Alom, M. Z., Taha, T. M., Yakopcic, C., Westberg, S., Sidike, P., Nasrin, M. S., ... & Asari, V. K. (2018). The history began from AlexNet: A comprehensive survey on deep learning approaches. arXiv preprint. arXiv:1803.01164.

Bharath, H. A., Chandan, M. C., & Nimish, G. (2019). Assessing land surface temperature and land use change through spatio-temporal analysis: A case study of select major cities of India. *Arabian Journal of Geosciences*, 12(11), 367. doi:10.1007/s12517-019-4547-1

Breunig, M., Bradley, P. E., Jahn, M., Kuper, P., Mazroob, N., Rösch, N., … & Jadidi, M. (2020). Geospatial data management research: Progress and future directions. *ISPRS International Journal of Geo-Information*, 9(2), 95. doi:10.3390/ijgi9020095

Gröger, G. H., Kolbe, T., & Czerwinski, A. (Eds.). (2006). Candidate OpenGIS® CityGML implementation specification. *Open Geospatial Consortium*. https://www.ogc.org/

Jeganathan, C., Roy, P. S., & Jha, M. N. (2011). Multi-objective spatial decision model for land use planning in a tourism district of India. *Journal of Environmental Informatics*, 17(1), 15–24. doi:10.3808/jei.201100182

Schmitt, M., Hughes, L. H., Qiu, C., & Zhu, X. X. (2019). SEN12MS: A curated dataset of georeferenced multi-spectral Sentinel-1/2 imagery for deep learning and data fusion. arXiv preprint. arXiv:1906.07789.

Syrris, V., Hasenohr, P., Delipetrev, B., Kotsev, A., Kempeneers, P., & Soille, P. (2019). Evaluation of the potential of convolutional neural networks and random forests for multi-class segmentation of Sentinel-2 imagery. *Remote Sensing*, 11(8), 907. doi:10.3390/rs11080907

2 Geospatial Big Data for Machine Learning

2.1 GEOSPATIAL BIG DATA

Spatial science, also referred to as geographic information science, plays a vital role in many scientific disciplines. It seeks to understand, analyse, and visualize real-world phenomena according to their locations. Look at our planet. It is not just continents and rivers. Its billion people live, work, and interact with their environment. The world has gone global; global economies, societies, and markets are interrelated and affect each other. By understanding these correlations, we can solve global challenges. So challenging problems have a scale like never before. Geospatial big data is a digital living inventory of the surface of our planet derived from five billion square kilometres of current and historical imagery and information. This inventory enables us to find, count, and measure the earth's surface features like never before. Let's try to understand this by an example of use case. Suppose a global retailer wants to open up a new store some place. First, they need to look at the existing shops around the globe, and check their successful percentages and reasons. Looking into the data means digging into their customer base, the type of neighbourhood, whether they have private land, how many cars they own, what is the square footage of their house, and then looking at the town as a whole. How many houses are there, how is the town experiencing growth, how many total personal vehicles are there, are the economic centres developing or moving from one place to another, etc. The retailer can look at the growth of the picture from neighbourhood to a town, to a state, or even a country. The analyser would start to get more information and ideas to make informed decisions. Geospatial big data has already enabled a number of industries and government applications to determine trends and correlations to answer the tough questions authorities are facing. The industry is excited with the advent of automated technologies such as machine learning and artificial intelligence to bring down monotonous data generation tasks from remote sensing images.

Big data is defined as high volume, high velocity, and high variety of data that can't be stored, managed, and processed by traditional tools. It requires a new way of storing, managing, and processing to enable insight discovery, decision-making, and process optimization (Beyer

DOI: 10.1201/9781003288046-2

and Laney, 2012). In the context of spatial data, we could say geospatial big data has high volume, velocity, and variety that exceeds the capability of current spatial computing platforms. Big data models are well explained by three Vs, i.e., volume, velocity, and variety. The volume corresponds to the enormous amount data generated every moment; velocity means data are growing and changing rapidly, and variety corresponds to data generated in multiple formats from multiple resources. This model defined by Beyer and Laney (2012) perfectly fits for geospatial datasets. At least 80% of all data are geographic, as most information around us can be georeferenced (Li et al., 2016). By this measure, 80% of the 2.5 exabytes (2,500,000,000 gigabytes) of big data generated every day are geographic.

The scientific field of geospatial artificial intelligence (GeoAI) is combining innovations in spatial science with the rapid growth methods in artificial intelligence (AI), particularly machine learning (e.g., deep learning), data mining, and high-performance computing to glean meaningful information from spatial big data. GeoAI is highly interdisciplinary, bridging many scientific fields, including computer science, engineering, statistics, and spatial science (VoPham et al., 2018). GeoAI is an emerging science that utilizes advances in high-performance computing to apply technologies in AI, particularly machine learning (e.g., deep learning) and data mining to extract meaningful information from spatial big data. GeoAI is both a specialized field within spatial science because particular spatial technologies, including GIS, must be used to process and analyse spatial data, and an applied type of spatial data science, as it is specifically focused on applying AI technologies to analyse spatial big data. GeoAI is the way to generate, process, and analyse spatial data by machines beyond normal human capabilities Because spatial data is more complex than traditional data, existing machine learning techniques are less effective at extracting spatial features or understanding a phenomenon or pattern. GeoAI combines spatial data mining, spatial machine learning and spatial statistics that are slightly eccentric to traditional techniques. We already know that plenty of industries and governments use big geospatial data; however, the automated tools for data creation have yet to become a full reality. The term GeoAI is most relevant in our times because of three reasons. Firstly, satellite and aerial sensors are generating a tremendous amount of data compared to the previous decade, and their projects will accelerate in the coming years. Secondly, machine learning models are maturing to deploy at scale and have started showing impressive outcomes. Lastly, cloud-based, server-based, and edge computing computation has revolutionized application development and deployment.

GeoAI is an emerging interdisciplinary scientific field that harnesses the innovations of spatial science, artificial intelligence (particularly machine learning and deep learning), data mining, and high-performance computing for knowledge discovery from spatial big data. GeoAI traces part of its roots from spatial data science, an evolving field that aims to help organize how we think about and approach processing and analysing spatial big data. Recent research demonstrates movement towards practical applications of GeoAI to address real-world problems from feature recognition to image enhancement.

2.2 MACHINE LEARNING FRAMEWORK FOR GEOSPATIAL BIG DATA

One motivation of building this new machine learning framework is to improve the geospatial big data generation process. Recent improvements in technology demand real-time spatial data processing and analytics and visualization to gain a completive advantage of real-time decision making. After carefully examining various scientific literature, there are various issues in geospatial big data processing and analysis. Therefore, this book presents new geospatial big data analytics and processing framework for geospatial data acquisition, data fusion, data storing, managing, processing, analysing, visualizing, and modelling. The purpose of spatial analysis is not only to identify a pattern but also to build models, if possible, by understanding the process. We believe that without a proper coordination and structuring framework, there will likely be much overlap and duplication amongst project phases, which can confuse each project participant's responsibilities.

A common mistake in geospatial big data projects is rushing into data collection and analysis, which prevents spending adequate time planning the amount of work involved in the project, understanding business requirements, or even defining the business problem properly. Geospatial big data is available all around us in various formats, shapes and sizes. Understanding the relevance of each of these datasets to business problems is a key aspect of success with the project. Also, geospatial big data has multiple layers of hidden complexity that are not visible by simply inspecting. Suppose a project does not identify the appropriate complexity and granularity level. The chances are high that an erroneous result set will occur that twists the expected analytical outputs. This book concentrates on developing a big data environment for geospatial data mining and machine learning. The data can be managed in the distributed environment to store enormous data. A big data environment for analysing geospatial data provides the ability to

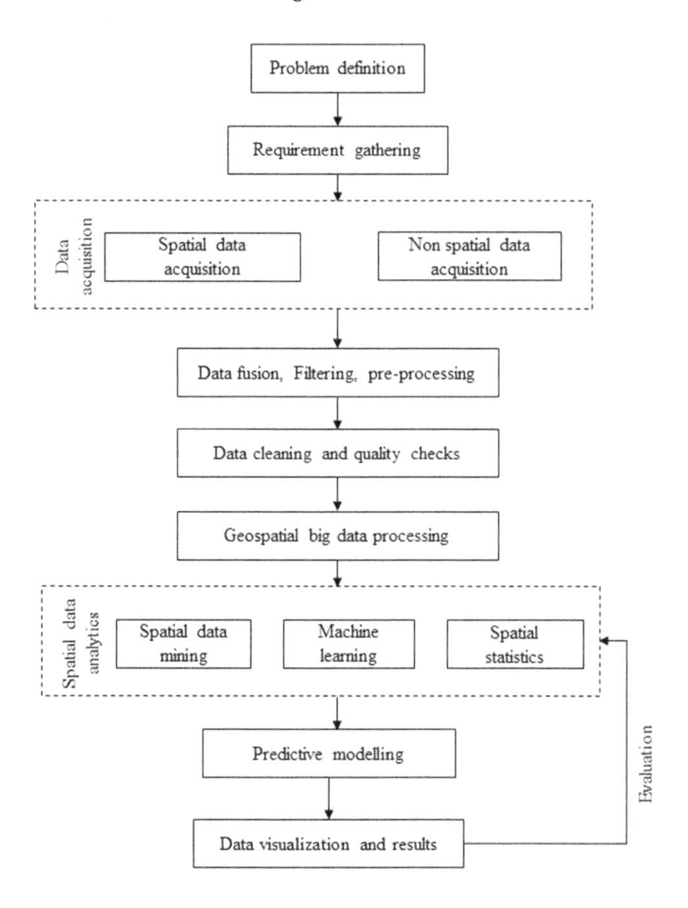

FIGURE 2.1 Machine learning framework for geospatial data.

deal with geospatial data on a large scale. A generic machine learning framework for geospatial datasets is shown in Figure 2.1.

The meaning of the two words, spatial data science and GeoAI, are always difficult to distinguish. Spatial science offers tools and technologies that enable us to understand, analyse, and visualize real-world phenomena according to their locations. It is a part of GeoAI in a larger context. VoPham et al. (2018) mentioned that as spatial data science continues to evolve as a discipline, spatial big data are constantly expanding, with two prominent examples being volunteered geographic information (VGI) and remote sensing. VGI encapsulates user-generated content with a locational component (Goodchild, 2007). Remote sensing is another type of spatial big data, which captures characteristics of objects from a distance, such as imagery from satellite sensors (Ma et al., 2015). This book concentrates on remote sensing-based big data capture and useful data generation using machine learning technologies.

2.3 DATA SOURCES

Depending on the sensor, remote sensing data can be expansive in both its geographic coverage (spanning the entire globe) as well as its temporal coverage (with frequent revisit times). We have seen an enormous increase in satellite remote sensing big data as private companies and governments continue to launch higher resolution satellites in recent years. For example, DigitalGlobe collects over 1 billion km^2 of high-resolution imagery each year as part of its constellation of commercial satellites, including the WorldView and GeoEye spacecraft. The U.S. Geological Survey and NASA Landsat program have continually launched earth-observing satellites since 1972, with spatial resolutions as fine as 15 m and increasing spectral resolution with each subsequent Landsat mission (e.g., Landsat 8 Operational Land Imager and Thermal Infrared Sensor launched in 2013 are composed of nine spectral bands and two thermal bands). The following section provides prominent earth observation datasets available at a global scale.

2.3.1 USGS – NASA's Mission

The United States Geological Survey (USGS) in collaboration with NASA has provided earth observation imagery data since 1972 to this day. These includes missions from Landsat 1 to Landsat 9, the recent Landsat 8 and Landsat 9 together, providing a full image of earth every eight days. The Landsat imagery datasets can be downloaded from Earth Explorer using a graphical user interface or landsatxplore Python package through a command-line interface or a Python API. A code snippet is provided here for downloading Landsat images using Python API, which can be used in Google Collaboratory.

Code snippet:

```
#installing libraries
pip install geopandas
pip install landsatxplore

#importing libraries
import pandas as pd
import geopandas as gpd
import os
import landsatxplore

#mount google drive
from google.colab import drive
drive.mount(/content/drive)
```

```
#set path to geojson file containing locations
for which images to be downloaded
path = path/name.geojson

#read geojson file using geopandas
df = gpd.read_file(path)

#get coordinates values for specified
locations
lat, long = df[latitude].values, df[longi
tude].values

#set Python API using Earth Explorer crede
ntials
from landsatxplore.api import API
api = API(username, password)

#create an empty list
image_list = []

#get Landsat scenes for specified date, cloud
cover, coordinates using API and for loop
for i in range(len(lat)):
 scenes = api.search(
    dataset=landsat_8_c1,
    latitude=lat[i],
    longitude=long[i],
    start_date=2021-01-01,
    end_date=2021-12-01,
    max_cloud_cover=50,
 )
# update the empty list with fetched Landsat
scenes
image_list.append(scenes)

#create an empty list
download_list = []

#create list of scenes with specified cloud
cover percentage
for i in range(len(image_list)):
 for j in range(len(image_list[i])):
  if (image_list[i][j][scene_cloud_cover]>= 3)
  & (image_list[i][j]
  [scene_cloud_cover] <= 10):
```

```
    download_list.append(image_list[i][j])
    break

#check the number of images to be downloaded
print(f{len(download_list)} scenes found for
download)

#Python API for image download using Earth
Explorer credentials
from landsatxplore.earthexplorer import Earth
Explorer
ee = Earth Explorer(username, password)

#downloading images using defined API and for
loop
for i in range(len(download_list)):
 ee.download(identifier=download_list[i]
 [landsat_scene_id], output_dir=path)

ee.logout()
```

2.3.2 Copernicus Missions

Copernicus is the European Union's earth observation programme, started its first satellite in 2014 and targeted to place about 20 satellites into orbit by 2030. The satellite constellation consists of the Sentinel family and other contributing missions. Atmosphere, Marine, Land, Climate change, Security and Emergency are six thematic areas to which Copernicus missions contribute. The datasets can be accessed through Copernicus Open Access Hub and also through Sentinel Python API.

Code snippet:

```
#import libraries
import sentinelsat
import os
import datetime
from sentinelsat import SentinelAPI, read_-
geojson, geojson_to_wkt

#set Python API using Copernicus credentials
api = SentinelAPI(username, password, https://
colhub.copernicus.eu/dhus/)

#set working directory containing multiple
geojson files with coordinates as bounds
```

```
os.chdir(path)

#create an empty list
geojson_files = []

#initialize n value for getting number of tiles
downloaded
n = 0

#read all geojson files into a list
for file in glob.glob(*.geojson):
  geojson_files.append(file)

#iterate through geojson files for downloading
one image for a location, for a given date in-
terval and cloud cover
for i in range(8,len(geojson_files)):
  footprint = geojson_to_wkt(read_geojson(geo
json_files[i]))
  products = api.query(footprint,
            date=(20211201, date(2021, 12, 31)),
            platformname=sentinel-2-l1c,
            cloudcoverpercentage=(45, 50))
  if len(products) != 0:
   single_image = products.popitem(last=False)
   api.download(single_image[0])
   n = n+1

#print the number of images downloaded to local
disc
print({} tiles downloaded.format(n))
```

2.3.3 ISRO Missions

The Indian Space Research Organization (ISRO) earth observation satellites have successfully established several operational applications in India and neighbouring nations. A wide number of users employ space-based inputs for varied reasons at both the federal and state levels. ISRO has launched Cartosat-1, 2, and 3; Resourcesat-1 and 2; Oceansat-1 and 2; Risat-1; Megha-Tropiques; SARAL; Scatsat; INSAT series; and a slew of other satellites. The National Remote Sensing Centre (NRSC) is a nodal entity that provides earth observation datasets from Indian satellites. Open datasets are available through the BHUVAN geoportal, while others can be purchased through the NRSC's dedicated ordering portal.

2.3.4 Other Missions

Apart from these earth observation missions, multiple prominent organizations worldwide are involved in developing and collecting earth surface datasets in numerous wavelengths of electromagnetic spectrum for a variety of applications. This includes both public domain or government agency satellites and commercial satellites. A comprehensive list of earth observation satellites can be found in the Wikipedia page. The application of machine learning is theoretically not limited to high-resolution optical images; instead, numerous use cases of satellite images can be found in the scientific literature (Ma et al., 2019). However, Table 2.1 gives a list of notable high-resolution satellite missions that can be used for building feature extraction through deep learning methods.

Apart from these high-resolution earth observation optical satellite images, researchers are exploring methods to extract fine details from

TABLE 2.1
List of Notable Earth Observation Satellites with Feature Extraction Capabilities

Name	Launch Year	Status	Resolution (Highest)
VRSS-1	2012	Active	2.5 m
Alsat-2A	2010	Active	2.5 m
Pleiades series	2001	Active	30 cm
Cartosat series	2005	Active	25 cm
Komsat series	1999	Active	70 cm
Khalifa sat (DubaiSat-3)	2013	Active	1.0 m
GeoEye-1	2008	Active	50 cm
PlanetScope	2016	Active	3 m
SkySat series	2013	Active	50 cm
SPOT series	1986	Active	1.5 m
TripleSat	2015	Active	80 cm
WorldView series	2007	Active	30 cm
IKONOS	2001	Inactive	1.0 m
Quickbird	2001	Inactive	65 cm
ALOS	2006	Inactive	2.5 m
Satellogic series	2013	Active	70 cm
Pelican series	2023–25	Future	50 cm
Albedo series	2024–25	Future	10 cm

earth surface features using a wide range of electromagnetic spectrum such as radiometric and microwave, Lidar remote sensing. The recent advancements in drone technology have opened a new avenue of data collection at even higher spatial resolution, in ranges of 1–5 cm.

2.4 THE CHALLENGE WITH EO DATA

Even though there are numerous sources that can identify how to get high-resolution satellite data, developing efficient automated techniques based on machine learning is a challenge mainly for two reasons. Firstly, the availability of labelled datasets concerning specific problem statements. The features on remote sensing imagery are vast and diverse, this creates a tremendous challenge to the feature extraction problem to generalize for large scale applications. Secondly, it needs heavy computational resources, requires a graphical processing unit with multi-core, multi-node facility for efficient and timely model training. To solve these problems and facilitate initial acceleration, several research and commercial organizations have taken the route of open challenge initiatives through hackathons or mapathons.

2.5 GeoAI platforms

A wide variety of software tools are already available to kickstart with machine learning for remote sensing datasets. The GeoAI platform ecosystem is expanding rapidly, with solutions from major cloud providers such as Google Cloud and AWS and a plethora of platforms that deliver services based on and utilizing remote sensing data. The Radiant Earth Foundation recently published a map of the earth observation market, identifying over 70 different organizations that offer spatial analytics. As a result, we could realize the vision of a fully resourced and equipped software cloud environment becoming a reality in the near future.

The GeoAI platforms available include application areas such as atmosphere, marine, climate change, security, emergency and land themes as a broader classification. However, here we shall focus on prominent GeoAI platforms that can be used for feature extraction from high-resolution remote sensing images. These platforms help to build machine learning workflows, training or building models, leveraging computation power of graphical processing units, testing, validation, deployment and even production level models can be built using existing platforms. Also, they provide some standard models for preliminary testing for common applications such as building footprint extraction, road extraction, and many others. The existing solutions are built from benchmark datasets that can be understood in their documentation. At the moment, fine feature extraction features from remote sensing images are at the proof-of-concept stage and improving rapidly. Table 2.2 provides a list of prominent GeoAI platforms that currently exist.

TABLE 2.2
Popular Platforms for Geospatial Data Machine Learning

Name	Focus Applications	Company	Commercial or Open Source	Top Features
Google Earth Engine	Multi-purpose	Google	Open	Planetary scale datasets, computational power, Python, JavaScript bindings
Cloud Native Geospatial	Multi-purpose	Linux Foundation	Open	Planetary scale geoprocessing, cloud-optimized GeoTIFF's, tiled web map view, crawlable metadata
Amazon Web Service	Multi-purpose	Amazon	Commercial	Planetary scale datasets, storage, computation power,
EOFactory	Agriculture, Infrastructure, Forestry	Skymap Global	Commercial	Change detection, cloud & haze removal, Image processing
Pictera	Infrastructure	Pictera (Swiss)	Commercial	GPU, API, pretrained models, training
Up42	Multi-purpose	Up42 (Germany)	Commercial	Data, algorithms, GPU, processing
ArcGIS online	Multi-purpose	ESRI	Commercial	Data, software tools, GPU, processing

2.6 CHOOSING THE RIGHT DATA

Datasets for deep learning model building must be selected depending on several factors. Much research is being done in the geospatial scientific community to address this challenge. The type of problem at hand, such as object detection or semantic segmentation, will determine what kind of images and labels will be employed in model construction. When selecting appropriate images and preparing labels, significant factors include type of geography, image characteristics, a feature of interest, computing capabilities, and application requirements. Let's consider extracting building footprints from satellite imagery of a city region for comprehension. If trained on similar surface conditions with accurate labels, existing algorithms may reasonably estimate building footprints (accuracy greater than 90%). Figure 2.2 shows a depiction of labelled images. It depicts a subset of the image with building footprints. The corresponding buildings are marked with polygons and transformed to a binary map, with the building representing pixel value '1' and the rest of the pixels as '0'. This type of labelled dataset must be prepared in order to train a machine learning model that can anticipate construction in new places. We took an image with specific characteristics (pixel size, bands, radiometry, and area of scene with buildings) and created an annotated image with similar characteristics. This dataset would only assist in the extraction of buildings, not other features such as roads, trees, lakes, parks, etc. Although the image may be identical, the ground truth or labelling must be based on the feature of interest when training the model.

A substantial number of annotated datasets are required to develop a deep learning model. Several organizations have made annotated datasets available in the open domain to facilitate additional deep learning research. Open domain deep learning datasets include ImageNet (Krizhevsky et al., 2017), the ISPRS benchmark dataset (Rottensteiner et al., 2012), and SpaceNet (Van Etten et al., 2018). In addition to high-resolution remote sensing images, open datasets give quality annotations. App developers, researchers, and start-ups benefit from open datasets. Open data sources are used in most deep learning research in remote

FIGURE 2.2 Buildings imaged and labelled with raster mask.

sensing. The dataset publisher often sponsors a competition where participants are asked to construct models. The following provide details on various free datasets and challenges.

UC Merced Land Use Dataset (Yang and Newsam, 2010)
The image databases, which contain 21 different land-use types, are available for research and development on the Internet. Each image is 256 × 256 pixels in size, with a spatial resolution of one foot. The satellite images were manually extracted for various cities around the country from large images in the USGS National Map Urban Area Imagery collection.

Northwestern Polytechnical University (Cheng et al., 2017)
This large-scale dataset aims to classify images with interclass differences in translation, viewpoint, pixel size, and target object. The datasets contain 31,500 remote sensing images divided into 45 classes, each containing 700 images.

Inria Aerial Image Labelling (Maggiori et al., 2017)
This dataset is for semantic segmentation of satellite images, focusing on building and non-building segmentation. Images of urban settlements from a range of cities are included in the dataset. Furthermore, the cities in the test set and the training set are different. This project aims to create a standardized framework for assessing classification approaches and, in particular, their generalization capacities.

Aerial Image Dataset (Xia et al., 2017)
The Aerial Image Dataset (AID) is a large-scale dataset used for aerial scene classification. The objective of AID is to advance state of the art in image scene classification using remote sensing. There are almost 10,000 annotated aerial scenes in the library. Higher interclass variability, smaller interclass dissimilarity, and huge size are the main characteristics of AID. The whole image set has 30 categories, most of which are urban features.

DeepGlobe (Demir et al., 2018)
The Satellite Image Understanding Challenge at DeepGlobe 2018 offers three public challenges for satellite segmentation, detection, and classification tasks. It shows how to overlay training images for road extraction, building identification, and land cover classification on satellite images using three datasets and evaluation techniques.

WHU-RS Dataset (Ji et al., 2018)
WHU-RS Dataset was created by hand using aerial and satellite imagery. The aerial dataset in Christchurch, New Zealand, has over 220,000

unique structures extracted from aerial images with a spatial resolution of 0.075 m and a coverage area of 450 km^2. Satellite imagery from QuickBird, Worldview series, IKONOS, and ZY-3 sensors has a ground resolution of 2.7 m and covers 550 km^2.

SpaceNet (Van Etten et al., 2018)
In August 2016, SpaceNet, an open innovation project, provided a collection of freely available imagery with co-registered map features. Before SpaceNet, computer vision researchers had not large enough labelled data for obtaining free, precision-labelled, high-resolution satellite images. SpaceNet collaborated with CosmiQ Works, Radiant Solutions, and NVIDIA to distribute a vast collection of annotated satellite images using Amazon Web Services (AWS) for scientific purposes. Building detection, road network detection, off-nadir building, road network extraction, and several more sophisticated challenges were among the SpaceNet challenges.

Data Fusion Contest (Le Saux et al., 2019)
Through a series of scientific challenges, the Image Analysis and Data Fusion Technical Committee of the Geoscience and Remote Sensing Society (GRSS) offers participants the opportunity to tackle remote sensing problems using multimodal data, leveraging new sensors and big data, and applying developing methodologies to extract geospatial information.

BigEarthNet (Sumbul et al., 2019)
BigEarthNet is a Sentinel-2 benchmark archive with multiple labels. The BigEarthNet is much larger than existing remote sensing (RS) archives, making it much easier to employ as a deep learning training technique. This data archive contains 590 Sentinel-2 tiles with resolutions of 120 × 120 pixels at 10 m resolution, 60 × 60 pixels at 20-m resolution, and 20 × 20 pixels at 20-m resolution, all of which have been annotated with multi-label categories.

SkyScapes (Azimi et al., 2019)
SkyScapes is a highly accurate and annotated at the pixel level for semantic segmentation remote sensing dataset produced from aerial surveys. The SkyScapes datasets provide annotations for 31 semantic categories and 12 subcategories, ranging from large objects including buildings, roads, lane markers, and vegetation. The primary purpose of this dataset is to do dense semantic segmentation as well as multi-class lane-marking prediction.

Agriculture-Vision (Chiu et al., 2020)
Agriculture-Vision is a large-scale aerial imagery resource for farmland. This is mostly for using semantic segmentation to determine agricultural

patterns. It includes 94,986 high-resolution images from 3,432 farmlands across the United States. The image has a 10-cm spatial resolution and RGB and NIR bands.

Aside from these datasets, there is an online collection of training-ready machine learning datasets. Table 2.3 lists some of the most well-known labelled datasets; however, it is neither exhaustive or inclusive of all available resources. Furthermore, a few platforms have made pre-trained models available for academics or users to experiment easily. A few well-known online resources are listed in Table 2.4 where pre-trained models can be found.

TABLE 2.3
Popular Labelled Remote Sensing Datasets for Training Machine Learning Models

Dataset	Format	Coverage	Current Status
Common Agricultural Practice (CAP) Austria dataset	Geopackage	National (Austria)	Legacy
Assess Building Damage (xView2)	Vector	Global	Legacy
Semantic segmentation of agricultural parcels from satellite time series PASTIS/PASTIS-R	Raster	France	New
Semantic segmentation – The MiniFrance Suite	Raster	France	New
Microsoft Building Footprints	Vector	Global	Legacy
Satellite image Classification Dataset-RSI-CB256	Raster	China	Legacy
Airbus Ship Detection	Raster	Global	Legacy
Airbus Aircraft Detection	Raster	Global	Legacy
Airbus Oil Storage Detection	Raster	Global	Legacy
NIST DSE Plant Identification with NEON Remote Sensing Data (IDTreeS data science challenge)	Raster, Vector	USA	Legacy
Cars Overhead with Context (COWC)	Raster, Vector	Toronto, Selwyn, Postdam, Vaihingen, Ohio, Utah	Legacy

(*Continued*)

TABLE 2.3 (Continued)
Popular Labelled Remote Sensing Datasets for Training Machine Learning Models

Dataset	Format	Coverage	Current Status
A Remote Sensing Land-Cover Dataset for Domain Adaptive Semantic Segmentation (LoveDa)	Raster, Vector	Nanjing, Changzhou, and Wuhan	Legacy
Sentinel-2 Cloud Mask Catalogue	Raster	Global	Legacy
LandCoverNet	Raster	Global	Legacy
Land Cover from Aerial Imagery (LandCover.ai)	Raster	Poland	Legacy
Cloud Segmentation on Satellite Images (95-Cloud)	Raster, Vector	Global	Legacy
Open AI Caribbean Challenge: Mapping Disaster Risk from Aerial Imagery	Raster	Caribbean	Legacy
Open Cities AI Challenge	Raster	Zanzibar	Legacy
SEN12MS	Raster	Global	Legacy
Sentinel-2 reference cloud masks generated by an active learning method	Raster, Vector	Unknown	Legacy
Agricultural Crop Cover Classification Challenge (CrowdANALYTIX)	Raster, Vector	USA	Legacy
DSTL Satellite Imagery Feature Detection Challenge	Raster	UK	Legacy
Spatial Procedures for Automated Removal of Cloud and Shadow (SPARCS) Validation Data	Raster	Global	Legacy
Biome: L8 Cloud Cover Validation data	Raster	Global	Legacy
Airbus Wind Turbines Patches	Raster	Unknown	Legacy
WiDS Datathon 2019: Detection of Oil Palm Plantations	Raster, Vector	Unknown	Legacy
So2Sat LCZ42	Raster	Global	Legacy
Cactus Aerial Photos	Raster	Mexico	Legacy
Planet: Understanding the Amazon from Space	Raster	Amazon	Legacy
Remote Sensing Image Scene Classification (RESISC)	Raster	Unknown	Legacy

TABLE 2.3 (Continued)
Popular Labelled Remote Sensing Datasets for Training Machine Learning Models

Dataset	Format	Coverage	Current Status
Deepsat: SAT-4/SAT-6 airborne datasets	Raster	USA	Legacy
Urban Semantic 3D (US3D) data	Raster	Jacksonville, Omaha	Legacy
Draper Satellite Image Chronology	Raster	California	Legacy
TiSeLaC: Time Series Land Cover Classification Challenge	Raster, Vector	France	Legacy
GHS built-up grid (GHS-BUILT-S2) – EC – Joint Research Centre	Raster	Global	New
Building Height – Copernicus	Raster	Europe	legacy
Land Surface Temperature – Copernicus	Raster	Global	New
Lake Surface Water Temperature – Copernicus	Raster	Global	New
Water Bodies – Copernicus	Raster	Global	New
GHS Urban Centre Database (GHS-UCDB) – EC – Joint Research Centre	Vector	Global	Legacy

TABLE 2.4
Machine Learning Models That Have Been Pre-Trained for a Variety of Applications

Pre-Trained Model	Description	Application	Current Status
BigEarthNet	BigEarthNet is a benchmark archive, consisting of 590,326 pairs of Sentinel-1 and Sentinel-2 image patches and pretrained models	Snow/cloud detection	Legacy

(*Continued*)

TABLE 2.4 (Continued)
Machine Learning Models That Have Been Pre-Trained for a Variety of Applications

Pre-Trained Model	Description	Application	Current Status
ArcGIS modes	Pretrained ArcGIS deep learning models remove the need for vast training data, massive compute resources, and advanced AI understanding. Increase productivity with built-in expertise and resources for feature extraction, land-cover categorization, image redaction, and object detection. Automate the analysis of images, point clouds, and video	Road extraction Building extraction Car detection Tree detection Land cover classification Settlement detection Parcel extraction Solar panel detection	New
Skeyenet	Identify and segment roads in aerial imagery	Road extraction	Legacy
Context-self contrastive pre-training	Semantic segmentation	Crop type segmentation	legacy
Inria aerial image challenge	Inria Aerial Image Labelling addresses a core topic in remote sensing: the automatic pixel wise labelling of aerial imagery	Building extraction	Legacy
Keras models	Keras Applications are deep learning models that are made available alongside pre-trained weights. These models can be used for prediction, feature extraction, and fine-tuning	Generic	Legacy
Microsoft	With Bing Maps, our AI-assisted mapping capabilities provides you with the most up-to-date building footprint data yet	Building extraction	Legacy
Radiant earth	Multiple datasets	Multiple	Legacy

REFERENCES

A Remote Sensing Land-Cover Dataset for Domain Adaptive Semantic Segmentation, https://github.com/Junjue-Wang/LoveDA
Agricultural Crop Cover Classification Challenge (CrowdANALYTIX), https://www.crowdanalytix.com/contests/agricultural-crop-cover-classification-challenge
Airbus Aircraft Detection, https://www.kaggle.com/airbusgeo/airbus-aircrafts-sample-dataset
Airbus Oil Storage Detection, https://www.kaggle.com/airbusgeo/airbus-oil-storage-detection-dataset
Airbus Ship Detection, https://www.kaggle.com/c/airbus-ship-detection/data
Airbus Wind Turbines Patches, https://www.kaggle.com/datasets/airbusgeo/airbus-wind-turbines-patches
Albedo Series, https://albedo.com/
ALOS, https://www.satimagingcorp.com/satellite-sensors/other-satellite-sensors/alos/
Alsat-2A, https://directory.eoportal.org/web/eoportal/satellite-missions/a/alsat-2
ArcGIS Models, https://livingatlas.arcgis.com/en/browse/?q=dlpk#d=1&q=dlpk
Assess Building Damage (xView2), https://xview2.org/
Azimi, S. M., Henry, C., Sommer, L., Schumann, A., & Vig, E. (2019). Skyscapes fine-grained semantic understanding of aerial scenes. In *Proceedings of the IEEE International Conference on Computer Vision*, pp. 7393–7403.
BigEarthNet, https://bigearth.net/#faq
Biome: L8 Cloud Cover Validation Data, https://landsat.usgs.gov/landsat-8-cloud-cover-assessment-validation-data#Barren
Building Height – Copernicus, https://land.copernicus.eu/local/urban-atlas/building-height-2012?tab=mapview
Beyer, M. A. & Laney, D. (2012). The importance of "big data": A definition. Gartner. G00235055.
Cactus Aerial Photos, https://www.kaggle.com/datasets/irvingvasquez/cactus-aerial-photos
Cars Overhead with Context, https://gdo152.llnl.gov/cowc/
Cartosat Series, https://www.sciencedirect.com/science/article/pii/B9780124095489103215?via%3Dihub
Cheng, G., Han, J., & Lu, X. (2017). Remote sensing image scene classification: Benchmark and state of the art. *Proceedings of the IEEE*, 105(10), 1865–1883. doi:10.1109/JPROC.2017.2675998.
Chiu, M. T., Xu, X., Wei, Y., Huang, Z., Schwing, A. G., Brunner, R., & Wilson, D. (2020). Agriculture-vision: A large aerial image database for agricultural pattern analysis. In *Proceedings of the IEEE/CVF Conference on Computer Vision and Pattern Recognition*, pp. 2828–2838.
Cloud Segmentation on Satellite Images (95-Cloud), https://www.kaggle.com/datasets/sorour/95cloud-cloud-segmentation-on-satellite-images
Common Agricultural Practice (CAP) Austria Dataset, https://www.data.gv.at/katalog/dataset/invekos-schlaege-oesterreich-2019/resource/662bff60-76a9-4fe4-90c9-9266b84959de

Context-self Contrastive Pretraining, https://github.com/michaeltrs/DeepSat Models?ref=pythonawesome.com

Copernicus Open Access Hub, https://scihub.copernicus.eu

Copernicus Program, https://www.copernicus.eu/en/about-copernicus/copernicus-detail

Crowdsourcing Deforestation in the Tropics during the Last Decade, http://pure.iiasa.ac.at/id/eprint/17539/

Deepsat: SAT-4/SAT-6 Airborne Datasets, https://csc.lsu.edu/~saikat/deepsat/

Demir, I., Koperski, K., Lindenbaum, D., Pang, G., Huang, J., Basu, S., … & Raska, R. (2018, June). Deepglobe 2018: A challenge to parse the earth through satellite images. In *2018 IEEE/CVF Conference on Computer Vision and Pattern Recognition Workshops*, pp. 172–17209. https://arxiv.org/abs/1805.06561v1

Draper Satellite Image Chronology, https://www.kaggle.com/c/draper-satellite-image-chronology

DSTL Satellite Imagery Feature Detection Challenge, https://www.kaggle.com/c/dstl-satellite-imagery-feature-detection

Earth Explorer, https://earthexplorer.usgs.gov

GeoEye-1, https://www.satimagingcorp.com/satellite-sensors/geoeye-1/

GHS Built-up Grid (GHS-BUILT-S2) - EC - Joint Research Centre, https://ghsl.jrc.ec.europa.eu/ghs_bu_s2_2018.php

GHS Urban Centre Database (GHS-UCDB) - EC - Joint Research Centre, https://ghsl.jrc.ec.europa.eu/ghs_stat_ucdb2015mt_r2019a.php

Goodchild, M. F. (2007). Citizens as sensors: The world of volunteered geography. *GeoJournal*, 69(4), 211–221.

IKONOS, https://microsites.maxar.com/ikonos20/#land-remote-sensing

Inria Aerial Image Challenge, https://project.inria.fr/aerialimagelabeling/leaderboard/

Ji, S., Wei, S., & Lu, M. (2018). Fully convolutional networks for multisource building extraction from an open aerial and satellite imagery data set. *IEEE Transactions on Geoscience and Remote Sensing*, 57(1), 574–586.

Keras Models, https://keras.io/api/applications/

Khalifa Sat (DubaiSat-3), https://directory.eoportal.org/web/eoportal/satellite-missions/k/khalifasat

Komsat Series, https://directory.eoportal.org/web/eoportal/satellite-missions/k/kompsat-3

Krizhevsky, A., Sutskever, I., & Hinton, G. E. (2017). ImageNet classification with deep convolutional neural networks. *Communications of the ACM*, 60(6), 84–90. doi:10.1145/3065386

Lake Surface Water Temperature – Copernicus, https://land.copernicus.eu/global/products/lswt

Land Cover from Aerial Imagery (LandCover.ai), https://landcover.ai/#dataset

Land Surface Temperature – Copernicus, https://land.copernicus.eu/global/products/lst

LandCoverNet, https://mlhub.earth/10.34911/rdnt.d2ce8i

Landsat Program, https://www.usgs.gov/landsat-missions/landsat-satellite-missions4

LandsatXplore, https://pypi.org/project/landsatxplore/

Le Saux, B., Yokoya, N., Hansch, R., Brown, M., & Hager, G. (2019). 2019 data fusion contest. *IEEE Geoscience and Remote Sensing Magazine*, 7(1), 103–105.

Li, S., Dragicevic, S., Castro, F. A., Sester, M., Winter, S., Coltekin, A., … & Cheng, T. (2016). Geospatial big data handling theory and methods: A review and research challenges. *ISPRS Journal of Photogrammetry and Remote Sensing*, 115, 119–133.

Ma, L., Liu, Y., Zhang, X., Ye, Y., Yin, G., & Johnson, B. A. (2019). Deep learning in remote sensing applications: A meta-analysis and review. *ISPRS Journal of Photogrammetry and Remote Sensing*, 152, 166–177.

Ma, Y., Wu, H., Wang, L., Huang, B., Ranjan, R., Zomaya, A., & Jie, W. (2015). Remote sensing big data computing: Challenges and opportunities. *Future Generation Computer Systems*, 51, 47–60.

Maggiori, E., Tarabalka, Y., Charpiat, G., & Alliez, P. (2017, July). Can semantic labelling methods generalize to any city? The Inria Aerial Image Labelling benchmark. In *2017 IEEE International Geoscience and Remote Sensing Symposium*, pp. 3226–3229. doi:10.1109/IGARSS.2017.8127684

Microsoft, https://www.microsoft.com/en-us/maps/building-footprints

Microsoft Building Footprints, https://www.microsoft.com/en-us/maps/building-footprints

NIST DSE Plant Identification with NEON Remote Sensing, https://www.ecodse.org/

Open AI Caribbean Challenge: Mapping Disaster Risk from Aerial Imagery, https://www.drivendata.org/competitions/58/disaster-response-roof-type/page/143/

Open Cities AI Challenge, https://www.drivendata.org/competitions/60/building-segmentation-disaster-resilience/page/150/

Pelican Series, https://www.planet.com/pulse/planet-introduces-new-high-resolution-pelican-satellites-and-fusion-with-sar/

Planet: Understanding the Amazon from Space, https://www.kaggle.com/c/planet-understanding-the-amazon-from-space/data

PlanetScope, https://docs.sentinel-hub.com/api/latest/data/planet-scope/

Pleiades Series, https://www.intelligence-airbusds.com/imagery/constellation/pleiades/

Quickbird, https://www.satimagingcorp.com/satellite-sensors/quickbird/

Radiant Earth Foundation, ML for EO Market Map, https://www.radiant.earth/infographic/ml-for-eo-market-map/

Radiant Earth, https://www.radiant.earth/mlhub/

Remote Sensing Image Scene Classification, https://www.tensorflow.org/datasets/catalog/resisc45

Rottensteiner, F., Sohn, G., Jung, J., Gerke, M., Baillard, C., Benitez, S., & Breitkopf, U. (2012). The ISPRS benchmark on urban object classification and 3D building reconstruction. *ISPRS Annals of the Photogrammetry,*

Remote Sensing and Spatial Information Sciences I-3 (2012),Nr. 1, 1(1), 293–298. doi:10.5194/isprsannals-i-3-293-2012
Satellite Image Classification Dataset-RSI-CB256, https://www.kaggle.com/sohelranaccselab/satellite-data
Satellogic Series, https://satellogic.com/technology/satellites/
Semantic Segmentation – The MiniFrance Suite, https://www.kaggle.com/javidtheimmortal/minifrance
Semantic Segmentation of Agricultural Parcels from Satellite Time Series PASTIS, https://github.com/VSainteuf/pastis-benchmark
SEN12MS, https://mediatum.ub.tum.de/1474000
Sentinel Python API, https://sentinelsat.readthedocs.io/en/stable/
Sentinel-2 Cloud Mask Catalogue, https://zenodo.org/record/4172871#.Yjiot-rP3ZR
Sentinel-2 Reference Cloud Masks Generated by an Active Learning Method, https://zenodo.org/record/1460961#.Yjipz-rP3ZT
Skeyenet, https://github.com/Paulymorphous/skeyenet
SkySat Series, https://www.satimagingcorp.com/satellite-sensors/skysat-1/
So2Sat LCZ42, https://mediatum.ub.tum.de/1454690
Spatial Procedures for Automated Removal of Cloud and Shadow (SPARCS) Validation Data, https://www.usgs.gov/landsat-missions/spatial-procedures-automated-removal-cloud-and-shadow-sparcs-validation-data
SPOT Series, https://www.satimagingcorp.com/satellite-sensors/spot-7/
Sumbul, G., Charfuelan, M., Demir, B., & Markl, V. (2019, July). BigEarthNet: A large-scale benchmark archive for remote sensing image understanding. In *IEEE International Geoscience and Remote Sensing Symposium*, pp. 5901–5904. doi:10.1109/IGARSS.2019.8900532
TiSeLaC: Time Series Land Cover Classification Challenge, https://sites.google.com/site/dinoienco/tiselac-time-series-land-cover-classification-challenge?authuser=0
TripleSat, https://www.satimagingcorp.com/satellite-sensors/triplesat-satellite/
Urban Semantic 3D (US3D) Data, https://ieee-dataport.org/open-access/data-fusion-contest-2019-dfc2019
Van Etten, A., Lindenbaum, D., & Bacastow, T. M. (2018). SpaceNet: A remote sensing dataset and challenge series. arXiv preprint. arXiv:1807.01232.
VoPham, T., Hart, J. E., Laden, F., & Chiang, Y. Y. (2018). Emerging trends in geospatial artificial intelligence (geoAI): Potential applications for environmental epidemiology. *Environmental Health*, 17(1), 1–6.
VRSS-1, https://directory.eoportal.org/web/eoportal/satellite-missions/v-w-x-y-z/vrss-1
Ward, J. S., & Barker, A. (2013). Undefined by data: A survey of big data definitions. arXiv preprint. arXiv:1309.5821.
Water Bodies – Copernicus, https://land.copernicus.eu/global/products/wb
WiDS Datathon 2019: Detection of Oil Palm Plantations, https://www.kaggle.com/c/widsdatathon2019/data
Wikipedia, List of Earth Observation Satellites, https://en.wikipedia.org/wiki/List_of_Earth_observation_satellites
WorldView Series, https://earth.esa.int/eogateway/missions/worldview

Xia, G. S., Hu, J., Hu, F., Shi, B., Bai, X., Zhong, Y., … & Lu, X. (2017). AID: A benchmark data set for performance evaluation of aerial scene classification. *IEEE Transactions on Geoscience and Remote Sensing*, 55(7), 3965–3981. doi:10.1109/TGRS.2017.2685945

Yang, Y., & Newsam, S. (2010, November). Bag-of-visual-words and spatial extensions for land-use classification. In *Proceedings of the 18th SIGSPATIAL International Conference on Advances in Geographic Information Systems*, pp. 270–279.

3 Spatial Feature Extraction

3.1 FEATURE EXTRACTION

Information contained in remote sensing imagery is enormous. If you have ever observed high-resolution aerial or satellite images, you can appreciate the fantastic levels of detail. Humans can visualize and grasp large amount of data and point out specific features. This data generation approach does not scale, as it is labour intensive. Moreover, extracting the needed information in a specific format for particular use cases from remote sensing images is quite challenging. The concept of computer vision technology is applied to image processing to extract information from images. The entire process can be divided into four computer vision streams. They are semantic segmentation, object detection, instance segmentation, and classification. In the context of images, it is critical to understand the fundamental differences between these four terminologies. The technique of dividing an image into a set of classes is known as semantic segmentation. Mapping of land use and land cover is a classic example of semantic segmentation. All pixels in an image are assigned to one of several classes, such as urban, water, vegetation, and others. Object detection uses an object's coordinates and bounding boxes in an image to identify it. A good illustration of an object detection technique is tree detection from an aerial image. Instance segmentation is a hybrid of semantic segmentation and object detection techniques. Object class and localization are used to classify all pixels during instance segmentation. In other words, semantic segmentation treats several objects belonging to the same category as a single class.

On the other hand, instance segmentation recognizes specific objects within each class. Classifying individual images into a category or label is known as classification. Because of the large geographical coverage of unique images in a scene, classification is the least used method in remote sensing images. Essentially, any of these techniques, or a combination of them, can be used to extract features from a remote sensing image.

3.2 MACHINE LEARNING MODELS

Automated feature extraction approaches are challenged by the complicated nature of spatial variation in remote sensing images.

DOI: 10.1201/9781003288046-3

Remotely sensed images are more difficult to comprehend due to abrupt tonal shifts, various land surface features, and topographical height variations. Several machine learning techniques have been investigated; unfortunately, the results are insufficient for accurate feature extraction. To segment image pixels into defined classes, methods such as maximum likelihood classification, decision tree, random forest, Naïve Bayes, support vector machine (SVM), and multi-layer perceptron (MLP) are examined. Based on training parameters, these models predict the class of each pixel. The primary goal of new method development is to improve accuracy and efficiency. The following section delves into the specifics of popular machine learning algorithms for extracting features from remote sensing images.

3.2.1 Maximum Likelihood Classifiers

Shackelford (2003) investigate how to classify buildings from impervious surfaces by combining an object-based approach with fuzzy classification. It mentions several other factors such as shape, neighbourhood, and spectral statistics that can be used to undertake object-based classification. To distinguish between different types of buildings, more work is needed. The object-based classifier for image segmentation delivers better results than per-pixel classification for feature extraction from high-resolution images, according to Myint et al. (2011). The authors emphasized the advantages of object-based classification over pixel-based approaches of parameter setting and iterations. According to the authors, the downside of the object-based strategy is that existing computation capabilities fail to process a significant volume of data, resulting in system failures.

The maximum likelihood classifier assumes that all classes in all bands have a normal distribution. The algorithm calculates a pixel's probability. The algorithm determines the likelihood of a pixel falling into a specific class. Because high-resolution images from complicated urban environments contradict this idea, this method's image segmentation falls well short of the required accuracy. The mathematical creation of discriminant functions for each individual pixel in the image is given in Equation 3.1. Individual pixels are assigned to a class with the highest probability.

$$g_i(x) = \ln p(\omega_i) - \frac{1}{2} \ln | \textstyle\sum_i | - \frac{1}{2}(x - m_i)^T \textstyle\sum_i^{-1} (x - m_i) \quad (3.1)$$

where i represents number of classes; x number of bands or dimension of the data; $p(\omega_i)$ the probability of class ω_i occurrence in the data;

Σ_i the determinant of the covariance matrix of the data belonging to a class ω_i; Σ_i^{-1} inverse matrix; and m_i mean vector.

3.2.2 Random Forest

Random forest algorithms for image classification were demonstrated by Horning (2010), who claims that they are easier to use and more tolerant of overfitting and outliers than conventional machine learning algorithms. Kumar et al. (2011) observed that using auxiliary layers such as NDVI, EVI, elevation, slope, and aspect in addition to basic band data improved classification accuracy. In urban, forested land and mountainous terrain, overall accuracies improved by 7.6%, 6.7%, and 10.8%, respectively. A random forest is a collection of decision tree classifiers that evaluates many outputs and predicts the outcome using the resultant's statistical mode value (Breiman, 2001). A decision tree classifier is a binary classifier that operates with a sequence of test questions and conditions arranged in a tree form.

Implementation of random forest classification in R: Segment a remote sensing image into buildings and non-building areas using marked ground truth as polygon features.

Code snippet:

```
#Importing libraries
library(raster)
library(rgdal)
library(RStoolbox)
library(randomForest)

#Setting memory limit, RAM allocation
memory.limit(size=)

#Setting directory location
setwd(folder path)

#Import image bands, RGB for example

##Individual bands
blue <- raster(file.choose())
green <- raster(file.choose())
red <- raster(file.choose())

##All bands, RGB
rs_image <- brick(file.choose())
```

```
# Stack bands
rs_image <- stack(B1,B2,B3) # add ndvi

#Assign names to the bands
names(rs_image) <- paste0(B,c(1:3))

#Read ground truth shapefiles
train.build <- readOGR(file.choose())
train.nonbuild <- readOGR(file.choose())

#Extract training pixels from image using
polygons as a dataframe
df_build <- extract(rs_image, train.build, df
= TRUE)
df_nonbuild <- extract(rs_image, train.non-
build, df = TRUE)

#Assign value 1 to buildings and 0 to non-
building class in a separate field called class
df_build$class <- 1
df_nonbuild$class <- 0

#Merge two dataframes as single dataframe
training.df <- rbind(df_build,df_nonbuild)

#Model training using Random Forest classifier
model_rf <- randomForest(class ~ B1+B2+B3, da-
ta=training.df, importance=TRUE, ntree=500)

#Prediction for entire image
prediction_nb <- predict(rs_image, model_rf,
na.rm=T)

#Wrtiting the prediction as GeoTIFF image to
the working directory
writeRaster(prediction_rf, folder path/clas-
sified_rf.tiff, format=GTiff)
```

3.2.3 Naïve Bayes

Naïve Bayes classifiers are used to learn automated models for segmentation and classification utilizing positive and negative examples for user-defined semantic land cover categories (Aksoy et al., 2005). A Bayesian classifier capable of identifying remotely detected images with a few training samples could give good results. The Naïve Bayes algorithm is based on the Bayes theorem, which states that the likelihood of event A is equal to the probability of event B, as shown in Equation 3.2. The assumption here is that A and B are unrelated to one another. Aksoy et al. (2005) provide the mathematical formulation of the classifier and the use case, as illustrated in Equation 3.2. Lv et al. (2017) used a Naïve Bayes classifier and high-resolution aerial images to map land cover. The assumption of Naïve Bayes that all features are independent is not true in high-resolution images.

$$P(A/B) = \frac{P(B/A)P(A)}{P(B)} \tag{3.2}$$

Implementation of Naïve Bayes classification in R: Segment a remote sensing image into buildings and non-building areas using marked ground truth as polygon features.

Code snippet:

```
#Importing libraries
library(raster)
library(rgdal)
library(RStoolbox)
library(naivebayes)

#Setting memory limit, RAM allocation
memory.limit(size=)

#Setting directory location
setwd(folder path)

#Import image bands, RGB for example

##Individual bands
blue <- raster(file.choose())
green <- raster(file.choose())
red <- raster(file.choose())

##All bands, RGB
rs_image <- brick(file.choose())

# Stack bands
rs_image <- stack(B1,B2,B3) # add ndvi

#Assign names to the bands
names(rs_image) <- paste0(B,c(1:3))
```

```
#Read ground truth shapefiles
train.build <- readOGR(file.choose())
train.nonbuild <- readOGR(file.choose())

#Extract training pixels from image using
polygons as a dataframe
df_build <- extract(rs_image, train.build, df
= TRUE)
df_nonbuild <- extract(rs_image, train.non-
build, df = TRUE)

#Assign value 1 to buildings and 0 to non-
building class in a separate field called class
df_build$class <- 1
df_nonbuild$class <- 0

#Merge two dataframes as single dataframe
training.df <- rbind(df_build,df_nonbuild)

#Model training using Naÿve bayed classifier
model_nb <- naive_bayes(training.df[,2:3],
training.df[,4], data = training.df)

#Prediction for entire image
prediction_nb <- predict(rs_image, model_nb,
na.rm=T)

#Wrtiting the prediction as GeoTIFF image to
the working directory
writeRaster(prediction_nb,folder path/clas-
sified_nb.tiff, format=GTiff)
```

3.2.4 The SVM

The segmentation-based SVM is utilized, and detected shadows are used to identify the buildings (Benarchid et al., 2013). The work is unique because it uses shadow and segmentation parameters to train the model. The results for rural, suburban, and urban areas are compared in the case study to determine the method's suitability. Quality assessment is done at all three case areas and object levels. The Hough transform effectively generated vectorization of rectangular or circular buildings (San and Turker, 2010). This strategy is used in sparsely populated urban areas. It also necessitates a dense urban setting. The method's success is determined by the accuracy with which a classifier recognizes a properly patched building. This approach does not necessitate setting any initial parameters (Dornaika et al., 2016). To train SVM classifiers to predict buildings, image descriptors are used. This task is used to extract the true contour of a building's rooftop from an aerial orthophoto. A green band is used to distinguish shadows thrown by trees, and changed shadows are employed as signatures for building extraction (Gao et al., 2018).

Similarly, a method for sampling bare ground is needed, and object-oriented segmentation is predicted to produce superior results. Aamir et al. (2019) used wavelet transformation to develop algorithms for extracting buildings from low-contact satellite images. Perceptual grouping is used to discover later structures. According to the experiments, the technique required far less time than earlier methods.

The SVM creates an optimal separating hyperplane (Vapnik, 1995). The SVM algorithm can handle both linearly and nonlinearly separable data points. If classes are not linearly separable in space, a slack variable is introduced, which penalizes each misclassified data point and seeks to reduce the total of slack variables. It also solves the non-linear decision boundary problem by transferring the original dimension points to a higher dimension. The distribution can be separated linearly using the transformation function (φ). The SVM-based classifiers have been utilized in remote sensing image categorization since they became popular (Dixon, 2008; Benarchid, 2013; Bramhe et al., 2020). Because this model is based on data that lie on hyperplanes, the model's success is determined on the training data. The mathematical formulations of the model are described in Foody et al. (2004).

$$\text{Considering a training sample } \{(W_i d_i)\} \text{ for } i = 0 \text{ to } N \tag{3.3}$$

where X_i is the input vector; N is training sample size; and d_i is +1 or −1.

Based on class belongingness, SVM categorizes classes by using hyperplanes

$$\begin{aligned} W_i^T X_i + b \geq 0 \text{ for } d_i = +1 \\ W_i^T X_i + b \leq 0 \text{ for } d_i = -1 \end{aligned} \tag{3.4}$$

where X_i is the input vector; W is adjustable weight; and b is parameter bias.

The above method is a linear classifier proposed by Vapnik (1995). The nonlinear method of SVM classification was introduced by Misra et al. (2018), in which vectors are mapped into higher dimensions of feature space. Kavzoglu and Colkesen (2009) and Yang (2011) showed the use of various kernels in SVM classification.

$$\text{Linear: } K(X,\ X_i) := X^T\ Xi \tag{3.5}$$

$$\text{Polynomial: } K(X,\ X_i) := (\gamma\ X^T\ X_i + r)\rho,\quad \gamma > 0 \tag{3.6}$$

$$\text{Radial basis function: } K(X,\ X_i): \exp(-\gamma\ \|X - X_i\|2),\ \gamma > 0 \tag{3.7}$$

$$\text{Sigmoid: } K(X,\ X_i)\tanh(\gamma\ X^T\ X_i + r) \tag{3.8}$$

where X represents input vector; X_i represents feature space vector; ρ represents degree of polynomial; r represents bias term, polynomial and sigmoid; and γ represents gamma term of kernels (polynomial, radial basis function and sigmoid).

Implementation of SVM classification in R: Segment a remote sensing image into buildings and non-building areas using marked ground truth as polygon features.

Code snippet:

```
#Importing libraries
library(raster)
library(rgdal)
library(kernlab)
library(e1071)
library(RStoolbox)

#Setting memory limit, RAM allocation
memory.limit(size=)

#Setting directory location
setwd(folder path)

#Import image bands, RGB for example

##Individual bands
blue <- raster(file.choose())
green <- raster(file.choose())
red <- raster(file.choose())

##All bands, RGB
rs_image <- brick(file.choose())

# Stack bands
rs_image <- stack(B1,B2,B3) # add ndvi
```

```
#Assign names to the bands
names(rs_image) <- paste0(B,c(1:3))

#Read ground truth shapefiles
train.build <- readOGR(file.choose())
train.nonbuild <- readOGR(file.choose())


#Extract training pixels from image using
polygons as a dataframe
df_build <- extract(rs_image, train.build, df
= TRUE)
df_nonbuild <- extract(rs_image, train.non-
build, df = TRUE)

#Assign value 1 to buildings and 0 to non-
building class in a separate field called class
df_build$class <- 1
df_nonbuild$class <- 0

#Merge two dataframes as single dataframe
training.df <- rbind(df_build,df_nonbuild)

#Model training using radial basis kernel
of SVM
svm_model <- svm(class ~ B1+B2+B3, data = trai-
ning.df, kernel=radial)

#Prediction for entire image
prediction_svm <- predict(rs_image, svm_
model, na.rm=T)

#Wrtiting the prediction as GeoTIFF image to
the working directory
writeRaster(prediction_svm,folder path/cla
ssified_svm.tiff, format=GTiff)
```

3.2.5 Neural Networks

Work by Mokhtarzade and Zoej (2007) explains the neural network structure required to extract features from imagery having red, green, and blue bands, and experimentation reveals the optimal number of neurons to utilize for the best results. A neural network's impact on input parameters is predicted in the study. To improve the accuracy, more edge and textural information should be used as per findings. An

MLP neural network is used for roads extraction from Google Earth images. The concept of generating training data and the training process in neural networks is explained by Kahraman et al. (2015). This approach achieves a 93.65% accuracy in extracting road elements from imagery. Puttinaovarat and Horkaew (2017) used image fusion to increase classification performance. To categorize imagery into urban and non-urban categories, machine learning algorithms such as ANN and SVM were utilized. The authors recommend that improvements be made using a heuristic machine learning approach based on the findings. Artificial neural networks are made up of densely interconnected basic units called neurons, whose design is inspired by the function and structure of the human brain. The MLP is the basic neural network structure shown in Figure 3.1. The MLP model has three layers: an input layer, a hidden layer, and an output layer, with each layer connected to the next layer either completely or randomly. Input layer neurons store various input properties such as training classes or imagery data bands. In satellite image classification, the output layer neurons represent land-use types.

Each neuron j in the hidden layer computes the sum of input X_i weighted by respective connection weight w_{ij} and calculates output y_i as a function of the sum.

$$y_i = f\left(\sum W_{ji} X_i\right) \tag{3.9}$$

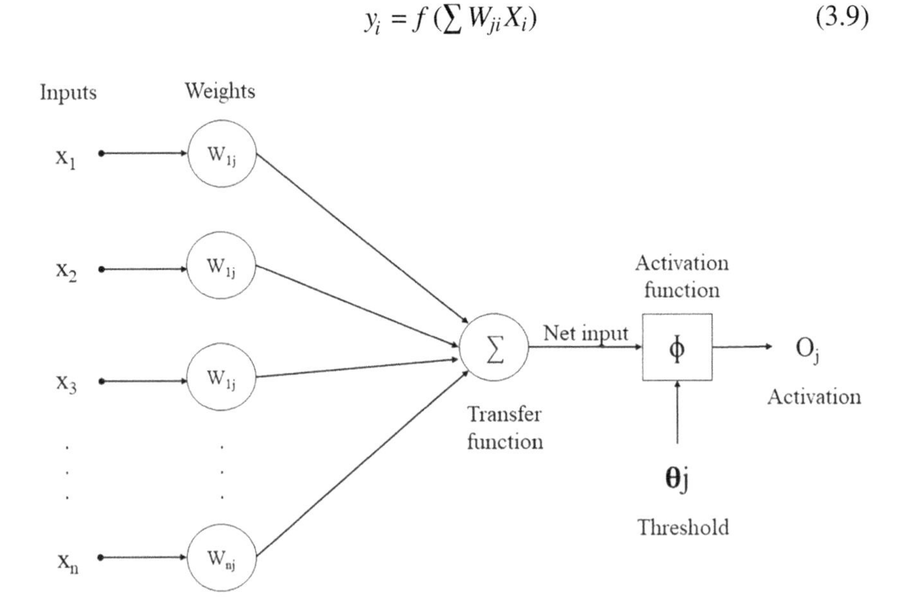

FIGURE 3.1 The node of MLP, the artificial neural network.

where f is an activation function that transforms the weighted sum of all inputs. Threshold function can be classified as a sigmoidal, radial basis function, rectified linear unit, etc. The sum of squared differences between the desired and actual values of output neuron e is given by

$$e = \sum_j (y_{dj} - y_j)^2 \tag{3.10}$$

The desired and actual value of output neuron j is y_{dj} and y_i, respectively. The model's effort is always to reduce the value of e by tuning the weights w_{ij}.

Implementation of neural network classification in R: Segment a remote sensing image into buildings and non-building areas using marked ground truth as polygon features.

```
#Importing libraries
library(raster)
library(rgdal)
library(nnet)
library(neuralnet
library(RStoolbox)

#Setting memory limit, RAM allocation
memory.limit(size=)

#Setting directory location
setwd(folder path)

#Import image bands, RGB for example

##Individual bands
blue <- raster(file.choose())
green <- raster(file.choose())
red <- raster(file.choose())

##All bands, RGB
rs_image <- brick(file.choose())

# Stack bands
rs_image <- stack(B1,B2,B3) # add ndvi

#Assign names to the bands
names(rs_image) <- paste0(B,c(1:3))
```

```
#Read ground truth shapefiles
train.build <- readOGR(file.choose())
train.nonbuild <- readOGR(file.choose())

#Extract training pixels from image using
polygons as a dataframe
df_build <- extract(rs_image, train.build, df
= TRUE)
df_nonbuild <- extract(rs_image, train.non-
build, df = TRUE)

#Assign value 1 to buildings and 0 to non-
building class in a separate field called class
df_build$class <- 1
df_nonbuild$class <- 0

#Merge two dataframes as single dataframe
training.df <- rbind(df_build,df_nonbuild)

#Create two data frames containing training
data or image band data and target values
values <- training.df[,2:3]
targets <- training.df[,4]

#Convertion to matrix
values <- as.matrix(values)
targets <- as.matrix(targets)

#Model training using neural network
nn_model <- nnet(values, targets, size=60,
rang = 0.1,decay = 5e-4,
linout=FALSE, maxit=300)

#Prediction for entire image
prediction_nn <- predict(rs_image, nn_model,
na.rm=T)

#Wrtiting the prediction as GeoTIFF image to
the working directory
writeRaster(prediction_nn,folder path/clas-
sified_nn.tiff, format=GTiff)
```

3.2.6 Convolutional Neural Networks

The most widely used deep learning model for image analysis is convolutional neural network (CNN), often known as ConvNet. The capacity of CNN to deduce spatial patterns from images is the most significant, making its application more efficient than any contemporary method. With the development of CNN-based image segmentation models, the performance of image segmentation models increased significantly. The CNN models employ a neural network structure in which hidden layer neurons are created as feature mappings using convolution. The spatial feature extraction can be separated into pre- and post-deep learning models due to their broad adoption and the huge jump in an accuracy improvement. CNN's success can be credited mainly to the models' improved grasp of images thanks to convolution filters. In several feature extraction challenge, many filters and deep networks yielded promising results. Patterns in the images, such as edges, shapes, textures, and objects, are detected using a variety of filters. The deeper convolutional layers could detect buildings, trees, waterbodies, vehicles, and other complex characteristics. Multiband images are fed into deep CNN models, which then pass them through convolution filters, pooling, and fully connected layers. The activation and loss functions are applied to specific layers inside the model. The following is a list of terms used in CNN models and their definitions:

Input layer: An input layer is a layer at the beginning of the process that contains an image in the form of a matrix, tensor, or array with a fixed width, height, and depth. Given the complexity of CNN and the current computational capacities utilized to train the model, the input layer size must be fixed (e.g., 512 × 512 or 256 × 256 pixels).

Convolution layers: A mathematical convolution procedure creates a convolutional layer from an input layer. The convolution layer is made up of several feature maps that are based on the operations using filters.

Convolution filters: Convolution filters, which have specific kernel sizes such as 7 × 7 × 3, are used to look at various parts of an image. The image dimension is reduced as a result of this procedure. The depth of the convolution filter must match the number of bands in the input image. These filters perform edge detection, image smoothing, object detection, classification, and other tasks.

Stride: The stride parameter controls the filter's movement across an image.

Padding: Padding is a CNN parameter that adds certain pixels around outer lines during convolution to prevent the image from shrinking.

Epoch: An epoch is a parameter that controls the number of times a neural network's weights are adjusted.

Batch size: Batch size refers to the number of images used in a single iteration (epoch) of a CNN training procedure. The batch size is determined by the compute capacity, network design, data volume, and objective. The batch size has a considerable influence on the model's accuracy.

Upsampling: Upsampling is a technique for increasing the resolution of feature maps in a CNN. The image is then rescaled to the required size, and the pixel values are determined using various interpolation algorithms. Deconvolution is another name for this approach.

Pooling layers: Pooling layers reduce the dimension of images by combining a fixed set of pixels to create a new contracted image. Pooling is done in a variety of ways, including maximum, minimum, and average. For example, a kernel of size 2×2 takes the largest value among pixels to produce a new image. The size of the kernel and the pooling procedure are employed depending on the reduction level required.

Fully connected layer: Every neuron in the previous and next layers is connected to every neuron in the fully connected layer. It is used to combine the images of feature maps from the previous layer into a single prediction or feature map. The fully connected layer is based on the standard MLP concept.

Activation functions: Each convolution layer has an activation function used to increase the output non-linearity. The output of a neuron is defined by its activation function given a set of inputs. Sigmoid, Rectified Linear Unit (ReLU), and other activation functions have been designed to execute a particular type of transition among neural networks.

Batch normalization: It is a technique for reducing the number of epochs required to train a model by stabilizing the learning process. The batch normalization approach normalizes outputs from the previous layer by subtracting the mean and then dividing the batch's standard deviation.

Dropout: It improves the efficiency of the learning process by removing a certain percentage of neurons from hidden layers of CNN. It introduces noise into the network in an indirect manner, allowing the model to be more generalized to deal with noise.

Loss function: The loss function in a neural network is the difference between the model's prediction and the actual values from the true labels. The goal of a neural network is to reduce the error or loss function to improve prediction accuracy. This is accomplished by solving optimization problems and utilizing an optimizer to minimize the loss function. Model learning refers to the process of reducing loss through optimization.

Optimizer: In a neural network, the optimizer's job is to minimize the difference between actual and predicted output in order to adjust the network's weights. They are in charge of modifying the learning rate and weights to obtain accurate results.

Learning rate: The learning rate is the step size taken when optimizing a loss function. The assigned learning rate might be anywhere from 0.01 to 0.0001. Repeated testing and tuning of a model determine the real value of the learning rate. A higher learning rate could result in overshooting or missing the real minimum. Taking a low number for the learning rate, on the other hand, takes a long time to minimize loss.

Training: CNN is a multi-layered neural network that uses back-propagation to train in a supervised manner. The definition of training a neural network is estimating the model's weights for a certain job. The term 'model training' is sometimes known as 'learning'. For example, multi-layer networks can learn low-, mid-, and high-level image information. CNN hierarchical feature's learning capacity allows it to learn a variety of image attributes quickly.

Transfer learning: This concept allows neural network learning at low to high levels to be transferred to a generic model. This makes it easier to retrain a model to generalize to different images. To maximize the outcomes, successive training of a neural network is used to alter the weights of targeted layers.

TensorFlow, PyTorch, Keras, and a few other open deep learning frameworks provide key tools for developing a CNN model. The developer must input images on various network layers with the appropriate parameters such as convolution filter size, strides, padding, activation function, pooling, and loss function. Hyperparameters include learning rate, stride, filter size, padding, and batch size, and utilized to fine-tune the model for optimal results.

Although a CNN model's basic structure remains intact, various CNN architecture is used among feature extraction tasks. The CNN architecture is primarily popular in analysing images and videos. Numerous models have developed and tested for different kinds of purposes. CNN architecture is developed and customized depending on the specific problem, such as image segmentation, classification, object detection, instant segmentation, etc. Previous studies and research projects have demonstrated efficient architecture for specific purposes encompassing concepts of CNN. The success of deep learning architecture can be attributed to the utilization of benchmark datasets, continuous improvement of parameters within the architecture, and high computational capabilities. The working principle of CNN-based deep learning architecture is explained in the subsequent section.

ConvNet, a deep learning module, is described in detail (Yuan, 2016) and is used to extract buildings in both urban and suburban areas. In a more complex and broader area than before, the trained network was effective. However, using a series of fixed-size sliding windows presented several complications. Zhang et al. (2016) proposed building detection in high-resolution images, the suggested method employs multi-scale saliency and CNN. When an overly dense building area is considered, precision drops from 89% to 78%. The satellite image is semantically segmented using a model called U-Net, which was originally created for medical image segmentation (Chhor et al., 2017). In this chapter, the authors used open street map data to train the model, demonstrating that the suggested approach can be learned with less labelled datasets than previous models. When utilizing nDSM to train FCN, improved accuracies are attained, and the resulting predictions are employed in CRF to produce superior binary building map outputs (Bittner et al., 2017). Tree cover or poor image representation are attributed for undetected buildings. In addition to nDSM, panchromatic or RGB images are predicted to improve the results. Building footprints were extracted from satellite imagery by Chawda et al. (2018) who then used a polygonization process to create building polygons. It will be fascinating to convert the building extraction challenge into an instance segmentation task. The authors employed heat maps and polygonization with marching squares in this study. Models such as FCN4, FCN8, and SegNet are utilized at a country scale to extract structures. According to the authors, a major problem for implementing the models is the limitation in GPU processing and the availability of ground truth labels (Yang et al., 2018).

3.3 DEEP LEARNING ARCHITECTURE

The applicability of remote sensing images has become more diverse because of recent advances in deep learning techniques and increased computational capacity. Building footprint extraction is an image segmentation task in the deep learning approach. Image segmentation is a method of categorizing all pixels in a digital image into a set of classes. Segmentation's main goal is to transform a raw image into a more meaningful representation in a specific context. Image segmentation (Fu et al., 2017) was done on satellite images using deep learning approaches to construct building footprints. CNN's components are aligned into several shapes to construct distinct architecture to achieve the most significant outcomes. CNN architecture can be classified according to their function or structure. Li et al. (2018) used the segmentation of pixels to classify numerous land-use classes and examined various deep learning approaches for image classification. Researchers worldwide have employed deep learning architecture including VGGNet, GoogleNet, ResNet, AlexnNet, SegNet, and U-Net to classify images into feature maps, with promising results. The following contains a collection of chosen satellite image segmentation studies:

U-Net (Ronneberger et al., 2015; Liu et al., 2019; Abdollahi et al., 2020)
Several research studies have employed the U-Net model based on encoder-decoder architecture previously established for biological-image segmentation to generate extraction from remote sensing images. Building extraction is suited to U-capacity to capture the context of spatial characteristics while maintaining precise localization. The U-Net design does not have a fully connected layer within and can be trained with a small number of training examples.

Seg-Net (Badrinarayanan et al., 2017; Yang et al., 2018)
A convolution filter is applied by an encoder in a Seg-Net architecture, followed by batch normalization, later non-linearity, and max-pooling. The high-frequency features are preserved, eliminating the need for up-sampling. Seg-Net has fewer parameters and uses less energy as a result of this. The problem with SegNet is that the gradient vanishes as the number of layers increases during the training process.

VGG-Net (Simonyan and Zisserman, 2014; Yang et al., 2018)
VGG-Net was proposed as a CNN design by Karen Simonyan and Andrew Zisserman of the University of Oxford in 2014. The impact of convolutional neural network depth on accuracy is the topic of this study. The number of parameters in the convolution layer and training time are reduced in VGGNet, resulting in reduced parameter training and time factor.

Res-Net (He et al., 2016; Wen et al., 2019)
Res-Net uses skip connections to tackle the problem of vanishing gradient. As a result, it saves a huge number of hidden layers from a

network. The skip connections are added to the VGG-19 network, which aids the identification of map features by introducing additional parameters.

Alex-Net (Lin et al., 2013; Vakalopoulou et al., 2015)
Alex-Net consists of a convolution layer, a pooling layer, normalization, convolution-pooling-normalization, a few more convolution layers, a pooling layer, and several fully connected layers. Five convolution layers and two completely connected layers before the last fully connected layer leads to the output classes.

Google-Net (Szegedy et al., 2015; Ostankovich and Afanasyev, 2018)
The Google-Net network was designed with computational efficiency and practicality in mind, allowing inference to be performed on single devices with limited computing power, especially those with limited memory. The network contains 22 layers when only layers with parameters are counted. The machine learning infrastructure system determines the number of layers required to form the network. Although there is an extra linear layer in this approach, average pooling is employed before the classifier.

DeepLab (Chen et al., 2017; Liu et al., 2019; Venugopal, 2020)
DeepLab is built with the network's speed, accuracy, and simplicity as top priorities. For semantic segmentation, the network employs atrous convolution and, in this example, PASCAL VOC 2012 datasets. The authors improve the findings by using atrous spatial pyramid pooling, which takes advantage of visual context at different scales.

U-Net (Ronneberger et al., 2015), a convolutional neural network (CNN)-based model developed for medical image segmentation, has lately been employed in many studies to extract features from satellite images (Chhor et al., 2017, Rastogi et al., 2020, Abdollahi, et al., 2020). This model structure can capture the context of spatial features and allow precise localization since it uses upsampling and downsampling techniques (Soni et al., 2020). Because this design uses fully convolutional layers as hidden layers, it can train the model with images of any size (McGlinchy et al., 2019). For semantic segmentation of satellite imagery into different land-use classes, Wu et al. (2019) used a modified U-Net architecture and a transfer learning technique. Xu et al. (2018) created Res-U-Net for building extraction and used guided filters to increase segmentation performance and establish a scope to include transfer learning with modified U-Net for building extraction.

The performance of deep learning models depends on the built-up conditions they are exposed to, as evidenced by careful monitoring of projected outputs from previous outcomes. This section of the book focuses on fine-tuning the model for an urban setting with various built-up conditions. The building footprints are extracted using a modified U-Net model in the following sections. The original U-Net model architecture (Ronneberger et al., 2015), previously designed to segment high-resolution aerial images, has been enhanced to suit the satellite image data and study area (Chawda et al., 2018). The model is trained utilizing transfer learning approaches because the study area has a variety of built-up environments, including commercial, sparse residential, dense residential, industrial, and mixed urban. In addition, standard deep learning architecture such as VGGNet, SegNet, and ResNet is used to provide a comparative examination of the model.

3.4 MODEL ARCHITECTURE

After learning about various model architecture, a prototype for image segmentation is created, including a contracting path and an expanding path known as the encoder and decoder. An encoder accepts a particular image size as input. It then generates higher-level features with reduced spatial resolution using a combination of convolutions and downsampling or pooling algorithms. The decoder's job is the opposite of the encoder's. The decoder performs upsampling and convolutions to generate the original image size from the contracted image. The compatibility of an encoder-decoder design with changeable image size or dimension is one of its advantages. Figure 3.2 depicts the architecture of the developed

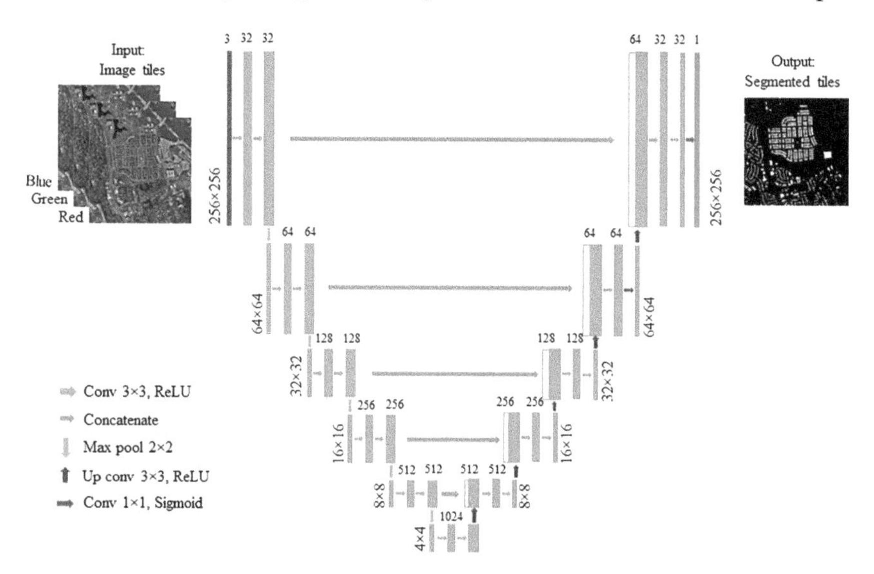

FIGURE 3.2 Modified U-Net model architecture.

model. The images are down-sampled, and convolutions are applied at each level, as seen on the figure's left side. This method is used to extract and understand the image's contextual information. The right side is utilized to upsample an image to its original size, which aids in feature localization. The model created can be broken into three components.

1. **Downsampling layers:** This has five blocks, with each block consisting of a 3 × 3 convolution layer, ReLU activation function 3 × 3 convolution layer, ReLU activation function, Max pooling, dropout, batch normalization. At the start of each block, the number of feature maps is doubled, from 32 to 512. The purpose of downsampling in this architecture is to find the contextual information existing in the image by employing skip connections to create an incremental path.
2. **Bottom line layer:** This layer consists of two convolution layers that lie between the two parts of the network model. This layer has 1024 feature maps.
3. **Upsampling layers:** The upsampling path comprises five blocks consisting of Concatenation layers, Batch normalization 3 × 3 convolution layer, ReLU activation function 3 × 3 convolution layer, ReLU activation function 3 × 3 up convolution or upsampling, dropout.

Finally, a 1 × 1 convolutional layer is used to translate 32 layers into a single layer with two classes and a sigmoid activation function. The predicted map is made up of probability ranging from 0 to 1, which must be translated to a binary map using a threshold that has been specifically chosen. A complete Python implementation of the model is given in Appendix 3.1 (https://docs.google.com/document/d/1q8XHFBgP3pTpfDl-JUhtt_PAsusiFSvY/edit?usp=sharing&ouid=109926032716705849211&rtpof=true&sd=true). The network's many parameters are detailed in the sections that follow.

3.4.1 Loss Function

A loss function called weighted binary cross-entropy is used in the current model. It's a cross-entropy version in which all of the coefficient weights are positive. The weighted binary cross-entropy reduces the biases in a dataset between the two classes. The loss function's capacity to overcome the problem of local minima improves when the weighted operation is used. As loss functions, intersection over union (IOU) and the Jaccard coefficient are used to create additional trails. The binary classification problem is said to be the reason why weighted binary cross-entropy performed better than other loss functions. The following is a description of the loss function's equation:

Let, P(Y=0) = p and P(Y=1) = 1–p and
(The predictions are from sigmoid function)

$$P(\hat{y} = 0) = \frac{1}{1 + e^{z}} = \hat{p} \text{ and } P(\hat{y} = 1) = 1 - \frac{1}{1 + e^{z}} = 1 - \hat{p} \tag{3.11}$$

Then, the loss function cross-entropy is given by

$$CE(p, \hat{p}) = -(\log(\hat{p}) + (1 - p)\log(1 - \hat{p})) \tag{3.12}$$

The variant called weighted cross-entropy is defined as

$$WCE((p, \hat{p}) = -(\beta \log(\hat{p}) + (1 - p)\log(1 - \hat{p}))) \tag{3.13}$$

Set $\beta > 1$ to decrease the number of false negatives.
Set $\beta > 1$ to decrease the number of false positives.

The sigmoid activation function is used at the last layer to predict probability values between 0 and 1. On a case-by-case basis, the value of can be adjusted based on the model's performance.

3.4.2 Data Augmentation

Data augmentation is a technique for efficiently increasing the variety of data used to train machine learning models. Cropping, flipping, padding, rotation, scaling, and other image manipulation operations expand the number of different datasets available for training models. Data augmentation was utilized for training the model by Marmanis et al. (2018) and Ji et al. (2019). In the case of insufficient training examples, data augmentation is utilized as a supplement dataset for various image segmentation problems. It aids in instilling the required invariance into the model during training. Due to changes in the spatial phenomena, image modification procedures such as rotate, flip, or any other tend to lose the spatial context of the features.

3.4.3 Hyperparameters

The model's hyperparameters had to be carefully set to match specific image properties. Learning rate, drop rate, batch size, image dimension, and the number of filters are all parameters to consider. The model takes longer to minimize the loss function if the learning rate is low, and if it is large, the step size may miss the minimum points. The drop rate is set to

0.2, which means that 20% of random neurons are removed from the network after each max-pooling layer for improved model performance. During model training, a batch size of 16 images with an image dimension of 256 × 256 pixels is employed for each epoch. As indicated in model representation (Figure 3.2), the number of filters or data layers changes in the model. Filters are doubled in each layer within the contracting path, whereas they are lowered by half in the expanding path.

3.4.4 Data Normalization

When the image dataset is normalized, the digital number (DN) is converted into a common scale. This is accomplished without distorting the datasets in terms of real DN values. By reducing the dataset's imbalances, this technique makes learning faster and more efficient. The normalizing formula is as follows:

$$\text{Normalized image} = (\text{Original image} - \text{Mean image}) / \text{Standard deviation}$$

3.4.5 Transfer Learning

The study's transfer learning technique involves repeatedly training the model for different build environments. The training datasets are collected from various locations throughout the city as a rectangular image measuring 5 × 5 square kilometres (km^2), and the model is trained many times. When visuals and instruction are combined simultaneously, the spatial context of the scene is lost. As a result, the model is trained using the transfer learning technique to avoid this loss. The trained model's weights are loaded, and subsequent iterative training is carried out by adjusting the weights. Transfer learning allows a single model to be trained for various built environments, allowing the model to be generalized.

3.5 METHODS

As shown in Figure 3.3, the overall method of extracting a building footprint is divided into four parts. The essential dataset is created in the first stage so that it can be used directly in the neural network model. The second stage, CNN-based deep learning models, is trained with images and masks using a transfer learning technique. In the third stage, the trained model extracts building footprints from an unknown area with different built-up environments. The final stage is the calculation of evaluation metrics utilizing ground truth or building masks.

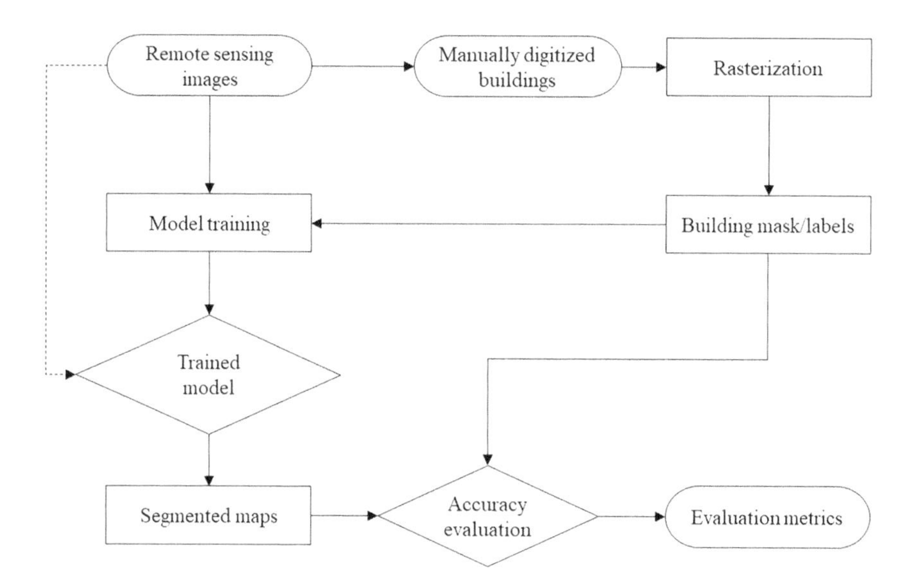

FIGURE 3.3 Flowchart depicting the general methodology for building feature extraction.

3.5.1 Image Pre-Processing

The image obtained by the TripleSat sensor from India's National Remote Sensing Centre (NRSC) was split into two parts: a panchromatic band with higher spatial resolution and multispectral bands with lesser spatial resolution. The panchromatic band is used to pan sharpen the multispectral image using the Intensity Hue Saturation (HIS) transformation for a higher Universal Image Quality Index (UIQI) value (Bharath et al., 2018). This provides higher spatial and spectral resolution. As a result, the resulting image comprises three bands from a multispectral image with improved spatial resolution. Building masks were created in raster format with the same metadata features as the source images, and the buildings were manually digitized. The image and masks are split into 256 × 256 tiles, with around 1,000 image tiles each. Before the training, these image tiles are normalized.

3.5.2 Model Training

The improved U-Net model offers a novel way to improve feature extraction from various built-up conditions. Eighty percent of datasets are utilized for training the model, and the remaining 20% are used to validate the model. While training, the accuracy of training and validation data is monitored, as shown in Figure 3.4. The performance of the model training is evaluated for a variety of input image sizes, including 64 × 64,

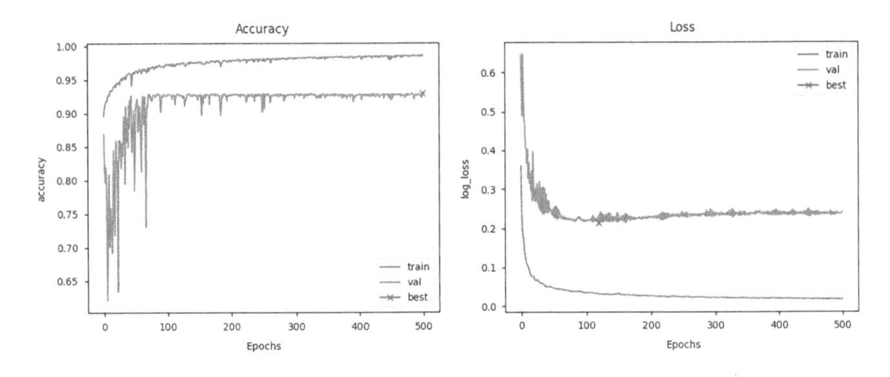

FIGURE 3.4 Model accuracy and model loss during training process.

96 × 96, 128 × 128, and 256 × 256, giving the computer system the ability to change batch sizes from 16 to 64.

3.5.3 Post-processing

Predictions are returned as a single band image with the same dimension as the input shape, which must be stitched together to make the complete image. The anticipated output image is given the coordinate reference system (CRS) and the datum of the original image. The model's prediction probability values range from 0 to 1. '0' denotes the background class, while '1' denotes the building class. A binary class thresholding method is utilized to convert the probabilistic output from the model values. The noise removal technique is used to remove unwanted artefacts in the output. Morphological closing and opening operators are employed to improve the accuracy of predictions or bring them closer to reality. Still, they often have the unintended consequence of lowering the output quality in terms of evaluation metrics.

3.5.4 Accuracy Evaluation

Extracting buildings from satellite images involves segmenting the image into two classes. As a result, it is defined as a classification problem, and the quality of the results is assessed using a confusion matrix (Shrestha and Vanneschi, 2018). The confusion matrix is a straightforward way to assess the segmentation algorithm's performance. It generates a variety of accuracy metrics, including classification accuracy, precision, recall, and F1-rate, among others. The classification accuracy ratio of successfully classified pixels to all classified pixels. On the other hand, precision is defined as the number

TABLE 3.1
Confusion Matrix and Accuracy Scores to Assess the Model's Performance

		Actual values		Accuracy = TP+TN/(TP+FP+FN+TN)
Predicted values		Positive	Negative	Precision = TP/(TP+FP)
	Positive	TP	FP	Recall = TP/(TP+FN)
	Negative	FN	TN	F1-rate = 2*Recall*Precision/(Recall + Precision)

of times a classifier is correct when predicting positive outcomes. The number of times a classifier predicted accurate values within a given series of positive values is used to evaluate its accuracy. The F1 rate is based on the harmonic mean, which penalizes extreme values. When precision and recall do not match, it aids in assessing both. Table 3.1 shows the formula for calculating accuracy parameters.

Calculating confusion matrix in R:

Code snippet:

```
#Importing libraries
library(raster)
library(rgdal)
library(caret)

#Set working directory
setwd(folder path)

# Import building mask and model prediction
mask <- raster(file.choose())
prediction <- raster(file.choose())

#Convert the images (mask & prediction) into
dataframe
x <- as.data.frame(mask)
y <- as.data.frame(prediction)

# Generate confusion matrix using R package
caret
cf  <-  confusionMatrix(factor(y$predicted_
commercial_mask),
factor(x$commercial_mask))
```

```
#Exporting confusion matrix into CSV file to
working directory

##Data clearning
cm_table <- as.matrix(cf)
cm_df <- as.data.frame(cm_table)
cm_df_all <- cbind(cm_df, fp=NA, fn=NA)
names(cm_df_all)[1]<-tp
names(cm_df_all)[2]<-tn
cm_df_all$fp[1] <- cm_df_all$tp[2]
cm_df_all$fn[1] <- cm_df_all$tn[2]
cm <- na.omit(cm_df_all)

#Writing confusion matrix as CSV file
write.csv(cm, file = cm.csv)
```

3.6 FINDINGS AND CONCLUSIONS

An 80-cm spatial resolution satellite image with RGB bands is used in the experiment. Images and binary building masks are used to train the designed models iteratively. The model is a trained transfer learning technique for various built-up environments, including 579 image tiles with a dimension of 256 × 256 and 167 images in the testing datasets. As the models are exposed to similar built-up conditions and trained thoroughly using heterogeneous samples, their ability to predict various types of buildings improves. Visual inspection of the model's output closely resembles the matching ground truth. The image and corresponding ground truth or building mask are given in Figures 3.5 and 3.6. A close examination of the results reveals that roads are identified as buildings that occasionally develop little undesirable patches. Several buildings with unusual shapes are also poorly captured. The model couldn't tell the difference between buildings that were close together. This underlines the model's inability to differentiate small elements in satellite images. Because images are split into 256 × 256 grids during prediction, there could be disjointed in developing prediction at the boundaries between two grids. This might result in the irregular development of building polygons during vectorization and the creation of holes in some circumstances. During the training stage, lower grid sizes such as 64 × 64, 96 × 96, and 128 × 128 were shown to increase the errors along these edges. The tendency could be the other way around, and it can be assessed with testing in more computing power. This model is intended for images with a rectangular shape.

Training the models to different urban built-up environments within the chosen city enhances the possibility of using transfer learning in

FIGURE 3.5 A satellite image used for training the model.

model building. The present experimental model is put to the test in built-up areas such as residential (sparse and dense), industrial, commercial, and mixed-use developments. The model is not exposed to any of these places during the training process. The model was tested for various built-up environments, which aided in the methodical knowledge of the forecast pattern. Finer projections for a residential area were clearly visible in the observation. On the other hand, commercial and industrial areas had fewer and more unique structures.

Residential buildings with similar shape, texture, and colour characteristics to the built-up residential condition were chosen for the training data. Even though the roofs with undetected colour elsewhere in the training data are large and separated by significant spacing, the model does not capture them correctly in the industrial built-up class. The improved U-Net outperformed other conventional models among developed models. The current experiment looked at Res-Net, Seg-Net, and VVG-Net.

The accuracy, precision, recall, and F1-score of the models are used to assess their performance (Table 3.2). The evaluation metrics demonstrate

FIGURE 3.6 Corresponding building mask used for training the model.

TABLE 3.2
Analysis of the Performance of Chosen Models

Models	Built-Up Conditions	Accuracy	Precision	Recall	F1-Score	RMSE
Modified U-Net	Sparse residential	0.95	0.98	0.96	0.97	2.00
	Dense residential	0.90	0.94	0.92	0.93	
	Commercial	0.95	0.98	0.96	0.97	
	Industrial	0.90	0.93	0.95	0.94	
	Mixed urban	0.90	0.97	0.91	0.94	
Res-Net	Sparse residential	0.93	0.97	0.96	0.96	2.63
	Dense residential	0.87	0.90	0.91	0.91	
	Commercial	0.93	0.96	0.95	0.96	
	Industrial	0.86	0.91	0.92	0.92	
	Mixed urban	0.87	0.93	0.91	0.92	

that when employing modified U-Net to segment an image in sparse residential built-up circumstances, an overall accuracy of 0.95 is attained. In comparison to the ground truth, the model accurately classifies 95% of the pixels in the image. The F1-score of 0.97 indicates that the obtained output is positively correlated and similar to the ground truth. Overall built-up conditions may be predicted with a minimum of 90% accuracy using the established model. In addition, when the RMSE of each model was calculated in terms of the area, it was discovered that U-Net performed better than the other models. The models are created using the Anaconda open-source software in a Python environment. The current model's training time was between 30 and 60 minutes. The manual digitizing of buildings took around two months for a 50-square-kilometre area with a density of 370 buildings per square kilometre.

REFERENCES

Aamir, M., Pu, Y. F., Rahman, Z., Tahir, M., Naeem, H., & Dai, Q. (2019). A framework for automatic building detection from low-contrast satellite images. *Symmetry*, 11(1), 3. doi:10.3390/sym11010003

Abdollahi, A., Pradhan, B., & Alamri, A. M. (2020). An ensemble architecture of deep convolutional Segnet and U-Net networks for building semantic segmentation from high-resolution aerial images. *Geocarto International*, 1–13. doi:10.1080/10106049.2020.1856199

Aksoy, S., Koperski, K., Tusk, C., Marchisio, G., & Tilton, J. C. (2005). Learning Bayesian classifiers for scene classification with a visual grammar. *IEEE Transactions on Geoscience and Remote Sensing*, 43(3), 581–589. doi:10.1109/TGRS.2004.839547

Badrinarayanan, V., Kendall, A., & Cipolla, R. (2017). SegNet: A deep convolutional encoder-decoder architecture for image segmentation. *IEEE Transactions on Pattern Analysis and Machine Intelligence*, 39(12), 2481–2495. doi:10.1109/TPAMI.2016.2644615

Benarchid, O., Raissouni, N., El Adib, S., Abbous, A., Azyat, A., Achhab, N. B., & Chahboun, A. (2013). Building extraction using object-based classification and shadow information in very high-resolution multispectral images, a case study: Tetuan, Morocco. *Canadian Journal on Image Processing and Computer Vision*, 4(1), 1–8.

Bharath, H. A., Vinay, S., Chandan, M. C., Gouri, B. A., & Ramachandra, T. V. (2018). Green to gray: Silicon Valley of India. *Journal of Environmental Management*, 206, 1287–1295. doi:10.1016/j.jenvman.2017.06.072

Bittner, K., Cui, S., & Reinartz, P. (2017). Building extraction from remote sensing data using fully convolutional networks. *International Archives of the Photogrammetry, Remote Sensing & Spatial Information Sciences*, 42. doi:10.5194/isprs-archives-XLII-1-W1-481-2017

Bramhe, V. S., Ghosh, S. K., & Garg, P. K. (2020). Extraction of built-up areas from Landsat-8 OLI data based on spectral-textural information and feature selection using support vector machine method. *Geocarto International*, 35(10), 1067–1087. doi:10.1080/10106049.2019.1566406

Breiman, L. (2001). Random forests. *Machine Learning*, 45(1), 5–32.

Chawda, C., Aghav, J., & Udar, S. (2018). Extracting building footprints from satellite images using convolutional neural networks. In *2018 International Conference on Advances in Computing, Communications and Informatics*, pp. 572–577. doi:10.1109/ICACCI.2018.8554893

Chen, L. C., Papandreou, G., Kokkinos, I., Murphy, K., & Yuille, A. L. (2017). Deeplab: Semantic image segmentation with deep convolutional nets, atrous convolution, and fully connected crfs. *IEEE Transactions on Pattern Analysis and Machine Intelligence*, 40(4), 834–848. doi:10.1109/TPAMI.2017.2699184

Chhor, G., Aramburu, C. B., & Bougdal-Lambert, I. (2017). Satellite image segmentation for building detection using U-Net. Web. http://cs229.stanford.edu/proj2017/final-reports/5243715.pdf

Dixon, B., & Candade, N. (2008). Multispectral landuse classification using neural networks and support vector machines: One or the other, or both? *International Journal of Remote Sensing*, 29(4), 1185–1206. doi:10.1080/01431160701294661

Dornaika, F., Moujahid, A., El Merabet, Y., & Ruichek, Y. (2016). Building detection from orthophotos using a machine learning approach: An empirical study on image segmentation and descriptors. *Expert Systems with Applications*, 58, 130–142. doi:10.1016/j.eswa.2016.03.024

Foody, G. M., & Mathur, A. (2004). A relative evaluation of multiclass image classification by support vector machines. *IEEE Transactions on Geoscience and Remote Sensing*, 42(6), 1335–1343. doi:10.1109/TGRS.2004.827257

Fu, G., Liu, C., Zhou, R., Sun, T., & Zhang, Q. (2017). Classification for high resolution remote sensing imagery using a fully convolutional network. *Remote Sensing*, 9(5), 498. doi:10.3390/rs9050498

Gao, X., Wang, M., Yang, Y., & Li, G. (2018). Building extraction from RGB VHR images using shifted shadow algorithm. *IEEE Access*, *6*, 22034–22045. doi:10.1109/ACCESS.2018.2819705

He, K., Zhang, X., Ren, S., & Sun, J. (2016, October). Identity mappings in deep residual networks. In *European Conference on Computer Vision*, pp. 630–645. doi:10.1007/978-3-319-46493-0_38

Horning, N. (2010). Random forests: An algorithm for image classification and generation of continuous fields data sets. In *Proceedings of the International Conference on Geoinformatics for Spatial Infrastructure Development in Earth and Allied Sciences,* Osaka, Japan, Vol. 911.

Ji, S., Wei, S., & Lu, M. (2019). A scale robust convolutional neural network for automatic building extraction from aerial and satellite imagery. *International journal of remote sensing*, 40(9), 3308–3322. 10.1080/01431161.2018.1528024

Kahraman, I., Turan, M. K., & Karas, I. R. (2015). Road detection from high satellite images using neural networks. *International Journal of Modeling and Optimization*, 5(4), 304. doi:10.7763/IJMO.2015.V5.479

Kavzoglu, T., & Colkesen, I. (2009). A kernel functions analysis for support vector machines for land cover classification. *International Journal of*

Applied Earth Observation and Geoinformation, 11(5), 352–359. 10.1016/j.jag.2009.06.002

Kumar, U., Dasgupta, A., Mukhopadhyay, C., & Ramachandra, T. V. (2011). Random forest algorithm with derived geographical layers for improved classification of remote sensing data. In *2011 Annual IEEE India Conference*, pp. 1–6. doi: 10.1109/INDCON.2011.6139382

Li, Y., Zhang, H., Xue, X., Jiang, Y., & Shen, Q. (2018). Deep learning for remote sensing image classification: A survey. *Wiley Interdisciplinary Reviews: Data Mining and Knowledge Discovery*, 8(6), e1264. doi:10.1002/widm.1264

Lin, M., Chen, Q., & Yan, S. (2013). Network in network. arXiv preprint. arXiv:1312.4400.

Liu, H., Luo, J., Huang, B., Hu, X., Sun, Y., Yang, Y., ... & Zhou, N. (2019). DE-Net: Deep encoding network for building extraction from high-resolution remote sensing imagery. *Remote Sensing*, 11(20), 2380. doi:10.3390/rs11202380

Lv, Z., Zhang, P., & Atli Benediktsson, J. (2017). Automatic object-oriented, spectral-spatial feature extraction driven by Tobler's first law of geography for very high-resolution aerial imagery classification. *Remote Sensing*, 9(3), 285. doi:10.3390/rs9030285

Marmanis, D., Schindler, K., Wegner, J. D., Galliani, S., Datcu, M., & Stilla, U. (2018). Classification with an edge: Improving semantic image segmentation with boundary detection. *ISPRS Journal of Photogrammetry and Remote Sensing*, 135, 158–172. doi:10.1016/j.isprsjprs.2017.11.009

McGlinchy, J., Johnson, B., Muller, B., Joseph, M., & Diaz, J. (2019). Application of U-Net fully convolutional neural network to impervious surface segmentation in urban environment from high resolution satellite imagery. In *IGARSS 2019.*

Misra, A., Vojinovic, Z., Ramakrishnan, B., Luijendijk, A., & Ranasinghe, R. (2018). Shallow water bathymetry mapping using support vector machine (SVM) technique and multispectral imagery. *International Journal of Remote Sensing*, 39(13), 4431–4450. doi:10.1080/01431161.2017.1421796

Mokhtarzade, M., & Zoej, M. V. (2007). Road detection from high-resolution satellite images using artificial neural networks. *International Journal of Applied Earth Observation and Geoinformation*, 9(1), 32–40. doi:10.1016/j.jag.2006.05.001

Myint, S. W., Gober, P., Brazel, A., Grossman-Clarke, S., & Weng, Q. (2011). Per-pixel vs. object-based classification of urban land cover extraction using high spatial resolution imagery. *Remote Sensing of Environment*, 115(5), 1145–1161.

Ostankovich, V., & Afanasyev, I. (2018, September). Illegal buildings detection from satellite images using Googlenet and cadastral map. In *2018 International Conference on Intelligent Systems*, pp. 616–623. doi:10.1109/IS.2018.8710565

Puttinaovarat, S., & Horkaew, P. (2017). Urban areas extraction from multi-sensor data based on machine learning and data fusion. *Pattern Recognition and Image Analysis*, 27(2), 326–337. doi:10.1134/S1054661816040131

Rastogi, K., Bodani, P., & Sharma, S. A. (2020). Automatic building footprint extraction from very high-resolution imagery using deep learning techniques. *Geocarto International*, 1–14. doi:10.1080/10106049.2020.1778100

Ronneberger, O., Fischer, P., & Brox, T. (2015, October). U-Net: Convolutional networks for biomedical image segmentation. In *International Conference on Medical Image Computing and Computer-Assisted Intervention*, pp. 234–241. Springer, Cham. doi:10.1007/978-3-319-24574-4_28

San, D. K., & Turker, M. (2010). Building extraction from high-resolution satellite images using Hough transform. *International Archives of the Photogrammetry, Remote Sensing and Spatial Information Science*, 38(8), 1063–1068.

Shackelford, A. K., & Davis, C. H. (2003). A combined fuzzy pixel-based and object-based approach for classification of high-resolution multispectral data over urban areas. *IEEE Transactions on GeoScience and Remote Sensing*, 41(10), 2354–2363. doi:10.1109/TGRS.2003.815972

Shrestha, S., & Vanneschi, L. (2018). Improved fully convolutional network with conditional random fields for building extraction. *Remote Sensing*, 10(7), 1135. doi:10.3390/rs10071135

Simonyan, K., & Zisserman, A. (2014). Very deep convolutional networks for large-scale image recognition. arXiv preprint. arXiv:1409.1556.

Soni, A., Koner, R., & Villuri, V. G. K. (2020). M-U-Net: Modified U-Net segmentation framework with satellite imagery. In *Proceedings of the Global AI Congress 2019*, pp. 47–59. Springer, Singapore. doi:10.1007/978-981-15-2188-1_4

Szegedy, C., Liu, W., Jia, Y., Sermanet, P., Reed, S., Anguelov, D., ... & Rabinovich, A. (2015). Going deeper with convolutions. In *Proceedings of the IEEE Conference on Computer Vision and Pattern Recognition*, pp. 1–9.

Vakalopoulou, M., Karantzalos, K., Komodakis, N., & Paragios, N. (2015, July). Building detection in very high-resolution multispectral data with deep learning features. In *IEEE International Geoscience and Remote Sensing Symposium*, pp. 1873–1876. doi:10.1109/IGARSS.2015.7326158.

Vapnik, V., Guyon, I., & Hastie, T. (1995). Support vector machines. *Machine Learning*, 20(3), 273–297.

Venugopal, N. (2020). Automatic semantic segmentation with DeepLab dilated learning network for change detection in remote sensing images. *Neural Processing Letters*, 1–23. doi:10.1007/s11063-019-10174-x

Wen, Q., Jiang, K., Wang, W., Liu, Q., Guo, Q., Li, L., & Wang, P. (2019). Automatic building extraction from Google Earth images under complex backgrounds based on deep instance segmentation network. *Sensors*, 19(2), 333. doi:10.3390/s19020333

Wu, M., Zhang, C., Liu, J., Zhou, L., & Li, X. (2019). Towards accurate high-resolution satellite image semantic segmentation. *IEEE Access*, 7, 55609–55619. doi:10.1109/ACCESS.2019.2913442

Xu, Y., Wu, L., Xie, Z., & Chen, Z. (2018). Building extraction in very high-resolution remote sensing imagery using deep learning and guided filters. *Remote Sensing*, 10(1), 144. doi:10.3390/rs10010144

Yang, H. L., Yuan, J., Lunga, D., Laverdiere, M., Rose, A., & Bhaduri, B. (2018). Building extraction at scale using convolutional neural network: Mapping of the United States. *IEEE Journal of Selected Topics in Applied Earth Observations and Remote Sensing*, 11(8), 2600–2614. doi:10.1109/JSTARS.2018.2835377

Yang, X. S., Deb, S., & Fong, S. (2011, July). Accelerated particle swarm optimization and support vector machine for business optimization and applications. In *International Conference on Networked Digital Technologies*, pp. 53–66. Springer, Berlin, Heidelberg. doi:10.1007/978-3-642-22185-9_6

Yuan, J., Yang, H. H. L., Omitaomu, O. A., & Bhaduri, B. L. (2016, December). Large-scale solar panel mapping from aerial images using deep convolutional networks. In *IEEE International Conference on Big Data*, pp. 2703–2708. doi:10.1109/BigData.2016.7840915

Zhang, Q., Wang, Y., Liu, Q., Liu, X., & Wang, W. (2016). CNN based suburban building detection using monocular high-resolution Google Earth images. In *IEEE International Geoscience and Remote Sensing Symposium*, pp. 661–664. doi:10.1109/IGARSS.2016.7729166

4 Building Height Estimation

4.1 SIGNIFICANCE OF BUILDING HEIGHT

A better comprehension of urban structure includes land-use patterns, built-up features, and building heights. For decades, multi-story or high-rise structures have accompanied the spread of metropolitan regions. This could be due to several factors, such as the high cost of land, the optimization of building construction costs, the kind of occupation, the rise in plinth and plot area ratio, etc. Because many of the structures that make up urban growth are multi-story, it is crucial to consider the third dimension while analysing them. City development officials choose housing flats and residential layouts since future construction must handle space optimization and vertical redevelopment. A cluster of high-rise apartments has replaced a slum of old buildings.

It is now necessary to capture the vertical growth of urban areas to evaluate the complete analysis of urban structures. Measurement of vertical urban expansion could aid urban planning, disaster management, environmental and ecological studies, energy evaluations, communications, resource allocation, and many other areas. Furthermore, the volume of built-up area in cities can be used to proxy for characteristics such as more significant economic activity. Using conventional approaches, data generation at the city scale is tedious, time-consuming, and labour-intensive. This makes it nearly impossible to collect data on urban buildings regularly, which is essential for successful decision-making.

Aerial surveys that take overlapping photographs using films aided in mapping 3D urban forms. Later, it was observed that digital photogrammetric approaches helped create the requisite datasets. Satellite stereoscopy has made significant breakthroughs in 3D mapping capabilities during the previous two decades. Even though drones can provide precise findings, the difficulties or limitations involved with flying operations and their expenditures remain significant obstacles. Satellite stereoscopic approaches are advantageous in cost, processing efficiency, and temporal possibilities for creating 3D maps of metropolitan structures at various scales. The resolution of the sensor, on the other hand, is what determines the output quality.

Traditional maps, which display earth characteristics on a two-dimensional surface, cannot convey the complexity of a three-dimensional

DOI: 10.1201/9781003288046-4

metropolitan setting. To accurately capture and display urban setup as it appears in the real world, 3D spatial datasets and powerful software tools are required. Remote sensing technology is being used to capture the shape and geometry of buildings. The most common are stereo satellite images, aerial photographs, and terrestrial data collection methods. Images can be produced using visible, infrared, microwave, and laser rays from the electromagnetic spectrum. This chapter focuses on the visible region of remote sensing with high-resolution imaging sensors.

Levels of detail (LODs) are used to categorize 3D city models based on the amount of visualization and analysis necessary. The Open Geospatial Consortium (OGC) has classified 3D models into five LODs (Gröger and Czerwinski, 2006). According to the specification, LOD1 contains well-defined heights of structures with flat roofs. This chapter focuses on the vertical components of the buildings and explains the context, process, and complications of preparing LOD1. In the next part, many studies relevant to building height estimations are explained in depth.

4.2 BACKGROUND

Monitoring urban growth, detecting illegal buildings, positioning telecommunication networks, measuring urban volumes, and estimating rooftop solar energy potential, disaster and risk hazards are a few potential applications of 3D building datasets. Multiple techniques, including laser and radar scanning, stereo satellite images, multi-angular imagery, and aerial surveys, are used to create 3D building datasets. Optical satellite stereo images have a cost and availability advantage over other local sensing systems such as laser and radar; nonetheless, output quality is a concern. Satellite stereo images are now accessible off the shelf for more extensive areas; faster processing times and higher resolutions make research and practical applications possible. Digital surface models (DSMs) are created by elevation models, including the above-ground objects depicted in Figure 4.1. The digital terrain model (DTM) is a raster model that simply represents the heights of the bare earth surface.

These elevation models are beneficial for research and practical applications, and various satellite agencies offer course resolution datasets for free. Free elevation models such as SRTM, ASTER, ALOS PRISM, Cartosat-30, and ALOS PALSAR are widely available and cater to many applications. They are, therefore, useful for large-scale data generation; however, their resolution remains limited at a finer scale. Another drawback is that these accessible datasets are only available for a limited time and do not provide continuous temporal coverage. However, using

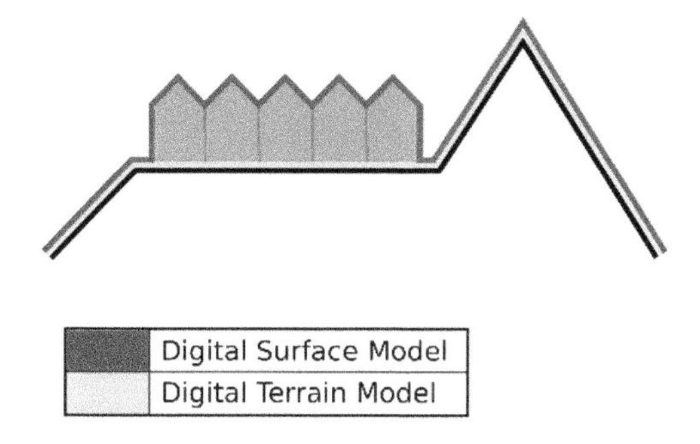

FIGURE 4.1 DSM and DTM representations.

freely available digital elevation models (DEMs) to calculate building heights has received less attention. Because of its temporal coverage capability, greater resolutions, and automated production process, commercially available satellite images are critical in preparing Digital Building Height (DBH). To generate depth information, satellite images captured with a stereo camera and multi-date images were used. In comparison to other commercially accessible datasets, Cartosat-1 is a low-cost stereo satellite image, as demonstrated in Table 4.1.

Several previous studies found that with accurate ground control points, Cartosat 1 satellite stereo images may produce DSMs with vertical accuracy of up to 4 pixels or 10 m. There are a variety of satellite

TABLE 4.1
Prominent Satellite Sensors with Stereo Imaging Capabilities

Satellite Sensor	Resolution		Swath in Kilometres	Available Since	Approx. Cost per Square Kilometre
	Pan	Multi-spectral			
SPOT 6/7	50 cm	50 cm	60	2014	$8.15
Geo-eye1	46 cm	1.84 m	15	2008	$55
KOMPSAT 3	40 cm	1.6 m	15	2012	$48
Pleiades-1B	50 cm	2.0 m	20	2012	$25
Pleiades-1A	50 cm	2.0 m	20	2011	$25
WorldView-2	46 cm	1.84 m	16	2009	$35
WorldView-3	31 cm	1.24 m	13	2014	$35
IKONOS	82 cm	3.28 m	11	1999–2006	$20
Cartosat-1	2.5 m	–	27	2005	$0.12
ALOS	2.5 m	10 m	35	2006	$0.37

sensors on the marketplace with stereo imaging capability for producing depth information. Several sensors with stereoscopic imaging capabilities were launched following IKONOS's successful initial attempts at stereo satellite images.

The stereoscopic parallaxes created by the roof and base of buildings on stereo models are used to quantify building height from satellite images manually. Automated algorithms use object shadows to calculate the height of structures from monocular images. The quality of detected shadows was impaired by taller nearby structures and spectral homogeneity of shadows with water. Unlike previous techniques, stereo satellite images that used to construct elevation models have gained widespread acceptance. Image matching techniques are widely employed in creating elevation models from stereo satellite images. DSMs obtained from stereo satellite images can be attributed to building height, tree height, and a variety of other height estimation use cases. Panagiotakis et al. (2018) found that adding multi-date image or tri-stereo images while modelling the elevation model improves its quality.

A bare earth model or DTM can be created from the surface model to calculate the height of above-ground items such as buildings or trees. The elevation of the earth's surface is represented by the DSM, including all above-ground objects. DTM, on the other hand, represents the elevation of the bare earth surface after all above-ground features have been removed. Arefi et al. (2011) devised an iterative approach for producing DTM using Cartosat-1 stereo images in their research. The DSM's filtering (Sithole and Vosselman, 2004; Zhang et al., 2016) is also used to get the closest height estimates possible. Ranagalage et al. (2018) used a grid generalization technique to create elevation grids all throughout the area. Beumier and Idrissa (2015) and Özcan et al. (2018) used segmentation to eliminate non-ground items before applying a region expanding algorithm to construct DTM.

Building heights are calculated using normalized DSM (nDSM = DSM – DTM) and a resampling technique described in work (Wurm et al., 2014), which yielded promising height estimates. Misra et al. (2018) uses a slope-based filtering technique to produce DTM from the current DSM to calculate building height.

4.3 ESTIMATION OF HEIGHT FROM STEREO SATELLITE IMAGES

This section explains how to use photogrammetric procedures to extract building heights from Cartosat-1 stereo images. The ability of the developed surface model to derive building heights is also addressed. Various terrain model generation methods are also briefly presented. Using

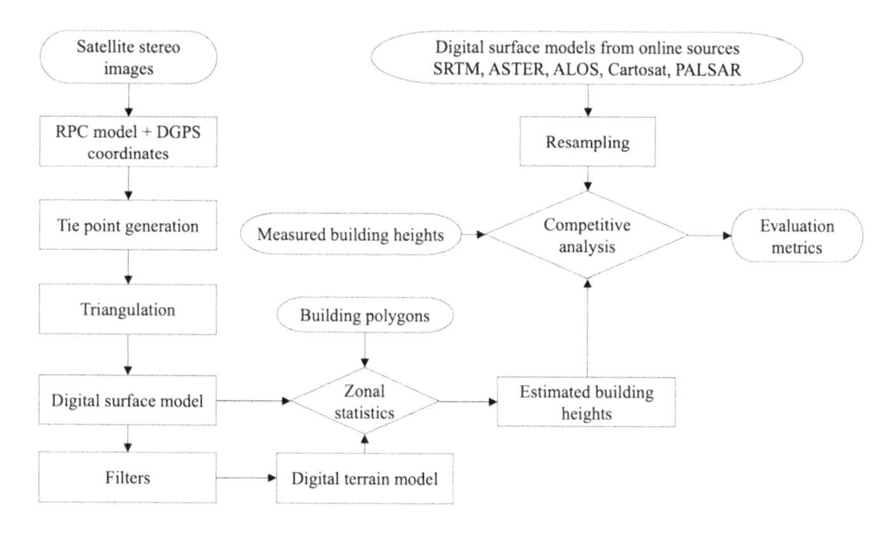

FIGURE 4.2 The flowchart showing the method adopted for building height estimation.

accurate GCP, the accuracy of generated DSM is assessed and compared to freely available elevation models. Figure 4.2 depicts the overall process for estimating building height using stereo satellite images. The following sections will go over the specifics of the methodology.

4.3.1 Stereo Satellite Images

DSM is generated using Cartosat-1 stereo satellite images with a spatial resolution of 2.5 m and compared to publicly available DEM from numerous sensors, as indicated in Table 4.2. The accuracy of the stereo model is increased by using 42 precise GCPs obtained throughout the study area. The 180+ different building heights measured manually are utilized to evaluate building height estimation from the model output, as shown in

TABLE 4.2
Spatial Datasets Used for Building Height Estimations

Sensor	Spatial Resolution (m)	Type	Capture
ALOS PRISM	30	DEM	April 2015
ALOS PALSAR	12.5	DEM	August 2015
Cartosat-1	30	DEM	April 2015
ASTER	30	DEM	2000–2009
SRTM	30	DEM	September 2014
Cartosat-1	2.5	Stereo images	March 2018

Appendix 4.2 (https://docs.google.com/document/d/1q8XHFBgP3pTpfDl-JUhtt_PAsusiFSvY/edit?usp=sharing&ouid=109926032716705849211&rtpof=true&sd=true). The required building polygons are manually digitized using QGIS software.

4.3.2 Surface Model Preparation

Stereo images and corresponding Rational Polynomial Coefficient (RPC) files for each image are included in the Cartosat-1 datasets. The RPC files are used to define the mathematical relationship between images at the moment of capture. To improve accuracy, the stereo pair of images is first orientated using RPC files, and then the model is refined using accurate GCPs. The RPC files are used to apply interior orientation to the stereo pair of images. Automatic tie points are generated afterward, and triangulation facilitates creating a more realistic model. The improved automatic terrain extraction (eATE, ERDAS Imagine) algorithm (Tsanis et al., 2014) is used for DSM creation. Figure 4.3 depicts

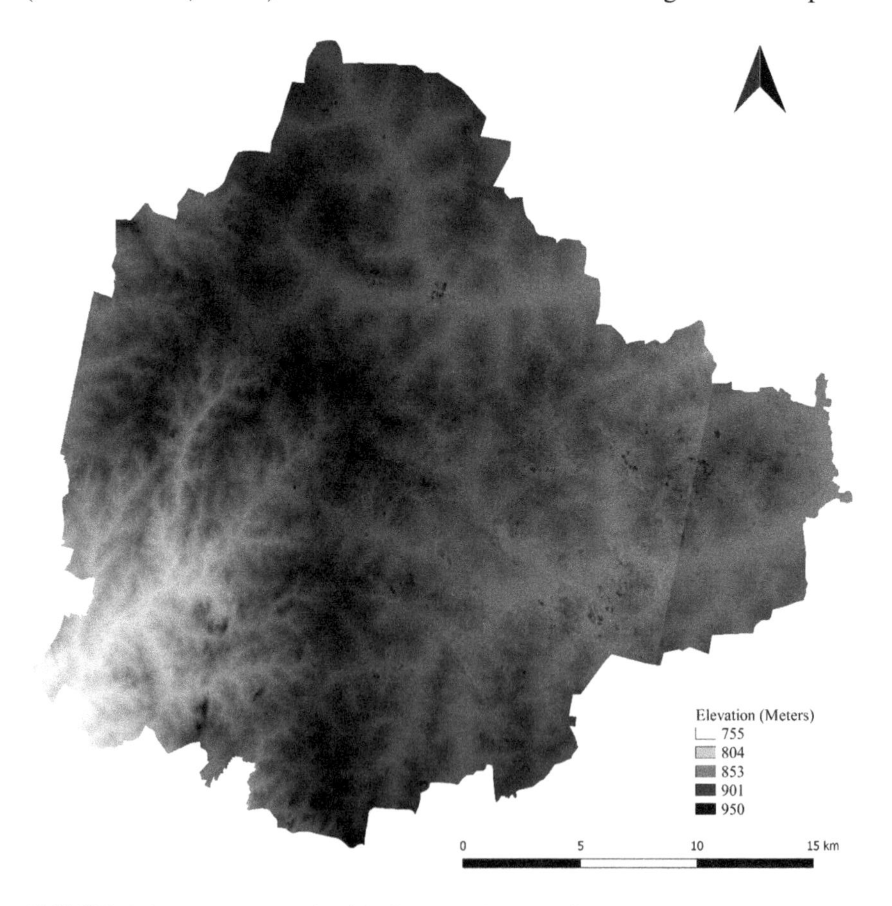

FIGURE 4.3 DSM created with Cartosat-1 stereo images.

the resulting DSM. The output cell size is limited to 5 m, which is twice the resolution of the original image.

4.3.3 DSM Quality Evaluation

The vertical accuracy of DSM derived from stereo images using the automated terrain extraction approach is examined. A comparison of the vertical accuracies of openly available DSMs is also carried out. Figure 4.4 depicts the GCPs that were utilized in the area. The control points are evenly distributed throughout the study region and were gathered via a differential GPS (DGPS) survey conducted in the area. The elevation corresponding to each location from DSMs is derived via overlay analysis. Then, using Equation 4.1, the root mean square error (RMSE) is determined independently for each DSM height.

$$\text{RMSE} = \sqrt{\frac{\Sigma((Z_{\text{DGPS}} - Z_{\text{DSM}}))^2}{N}} \tag{4.1}$$

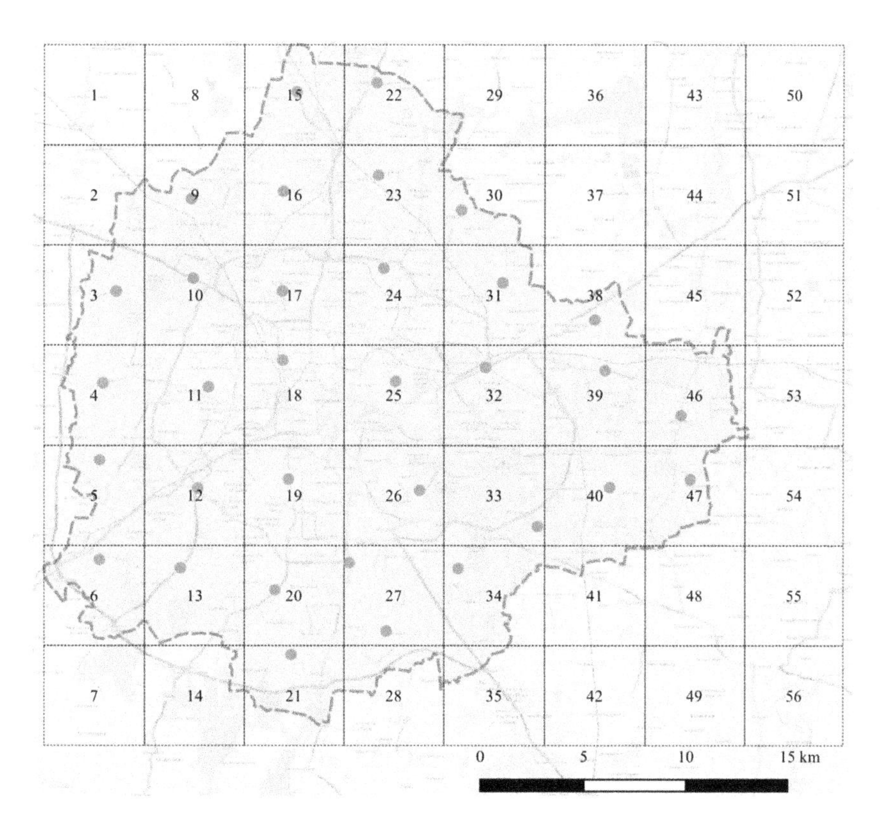

FIGURE 4.4 DGPS point locations across the city.

Z_{DGPS} corresponds to the elevation of points measured using DGPS and Z_{DSM} is the corresponding elevation obtained from DSM.

4.3.4 Preparation of a Terrain Model

Because of the diverse nature of the urban environment, creating an elevation model that excludes above-earth objects is a complex undertaking. It is challenging to generalize a single strategy for generating a terrain model because the metropolitan region has a variety of land-use categories spread out arbitrarily. Earlier work demonstrated automated methods such as segmentation, grid-based methods, multi-directional slope (MDS) dependent filtering, geodesic dilation, and others. A suitable method must be devised based on the terrain complexity and data type used for surface generation. In general, slope-based filtering outperforms other strategies among the procedures mentioned here. This part aims to assess the results of various ways and demonstrate the quality of the results produced by each approach.

To generate the DTM, five strategies are tested, and the resulting accuracy is assessed. The methods used are described in the sections that follow.

4.3.4.1 MDS Filtering

The MSD filtering approach was applied to high-resolution DSM on undulating terrain, and good results were obtained (Mousa et al., 2017). Researchers also used this method to construct a terrain model from low-resolution DSM (Misra et al., 2018). The MDS technique necessitates the adjustment of parameters such as the height threshold, slope threshold, scanline filter extent, and Gaussian smoothing kernel size. Each pixel is categorized as terrain or not using this technique, which involves assessing the height in five directions. It is considered a non-ground pixel if the height difference exceeds the set threshold. This method resulted in a terrain model with holes corresponding to non-ground objects. The holes are then filled with the closest elevation values using linear interpolation techniques. A flowchart depicting the MDS method is shown in Figure 4.5.

Here, S_t is the slope threshold, and Z_t is the height threshold given as input parameters.

4.3.4.2 Grid-Based Method

There are two steps to this method. The first phase determines the minimum elevation value in a grid of defined size (e.g., 50, 70, 100, 120 m). An overlay procedure of DSM and grid polygon is used to determine the minimum value for each grid identification. Each pixel under

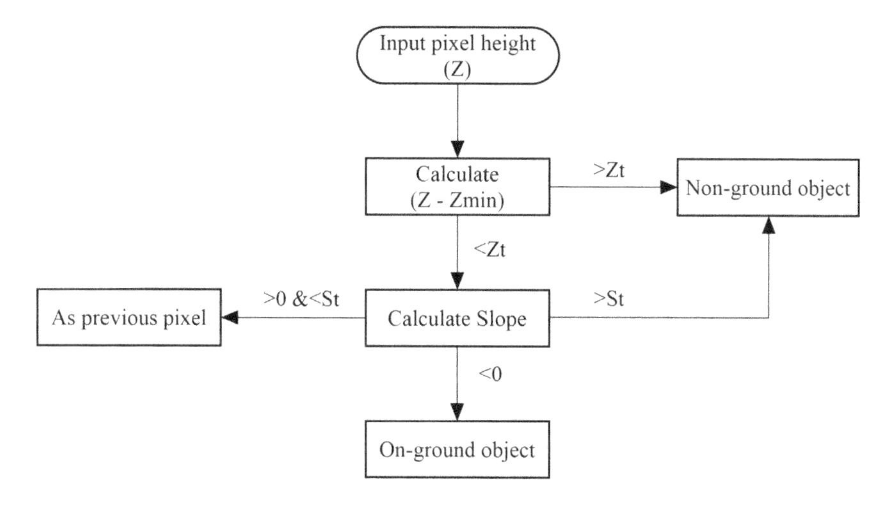

FIGURE 4.5 Flowchart illustrating the procedure for obtaining DTM from DSM.

that grid is then given a minimum elevation value. The disadvantage of this method is that it causes discontinuity at the grid's boundaries, resulting in inaccurate height estimates in those areas. Experiments revealed that a grid size of 100 m yielded better results than others.

4.3.4.3 Interpolation

In this method, the lowest elevation value in a grid is first found and mapped to the grid's centre. The Inverse Distance Weight (IDW) approach is then used to interpolate the minimal value to the full area. This process is repeated throughout the study region to fill the elevation values. Various grid spacing values, such as 50, 70, 90, and 120 m, are investigated in this chapter, with the 100-m value being chosen for superior outcomes.

4.3.4.4 Slope-Based Filter

This method has already been used to generate DTM using laser point cloud data (Vosselman, 2000; Asal, 2019). In grayscale imagery, this procedure is identical to the morphological erosion operation. When the filter runs across the image, it removes all above-ground items such as houses and trees, resulting in a bare earth surface. Later, the holes should be filled with elevation data using linear interpolation.

4.3.4.5 Road Buffers

This simple technique extracts elevation data from roads in the study region and then extrapolates them to the remaining areas. This is because roads represent ground elevations throughout the scene, and all building heights are measured from the road's level. To do this, the road layer is first applied to the whole study area, followed by a 2.5-m buffer on both sides of the road centreline. All elevation data under the road buffer are clipped and recorded as a distinct raster layer by using overlay analysis. The area belonging to non-road pixels is then filled with elevation data using linear interpolation.

4.4 ESTIMATING THE HEIGHT OF A BUILDING

Two elevation models, DTM and DSM, are required to estimate building height and polygons. Manual digitization produces the necessary building polygons to determine height parameters. As stated below, overlay analysis is used to determine the height of buildings.

4.4.1 DTM Method

The height map normalized DSM is obtained by subtracting DTM from DSM, as shown in Equation 4.2. The height of above-ground objects such as trees and buildings are stored in the nDSM. The height corresponding to the greatest value is determined via overlay analysis or zonal statistics tool for each building polygon. The height of each building is obtained from the maximum elevation value from nDSM within the building polygon as given in Equation 4.3.

$$\text{nDSM} = \text{DSM} - \text{DTM} \tag{4.2}$$

$$\text{Height of a building}_{(\text{polygon})} = \text{maximum}(\text{nDSM}_{\text{polygon}}) \tag{4.3}$$

4.4.2 Buffer Polygons

A buffer area is generated surrounding each building to a fixed distance of 25 m in this method, which was previously employed by Sharma et al. (2016). The ground elevation is then calculated using the DSM and a pixel corresponding to the minimum value in that buffer zone. The roof elevation is the maximum pixel value within the building polygon. Finally, the difference between the two-elevation data is used to calculate the approximate height of the building. Figure 4.6 depicts a visual representation of the building, buffer area, and DSM overlaid over images. Note that a terrain model is not required for this procedure; instead, the minimum and highest elevation values in the vicinity of a building from surface models can be used to estimate height.

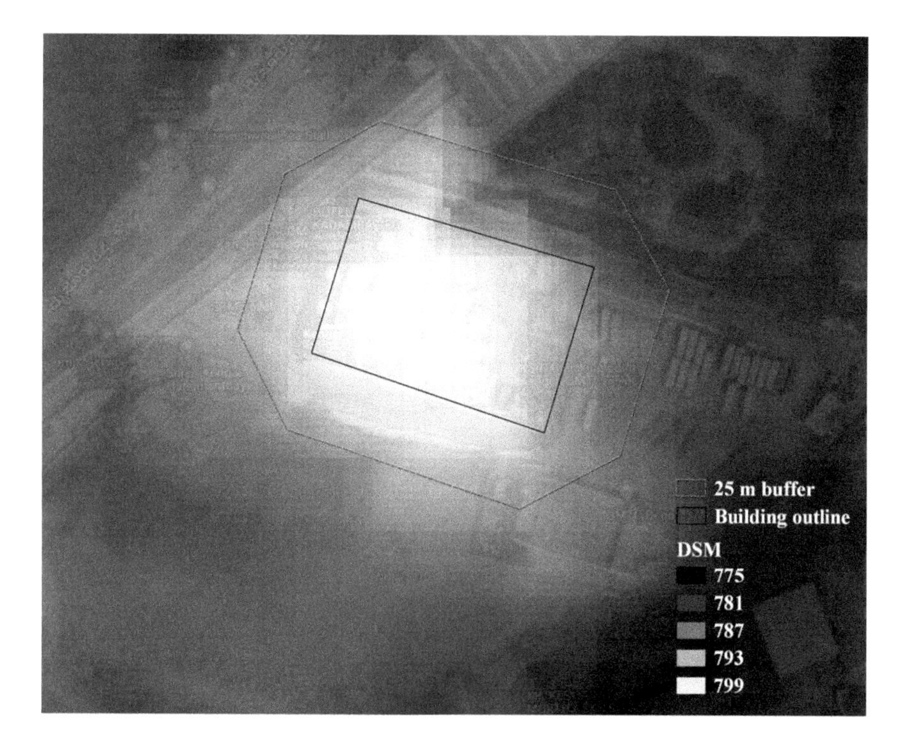

FIGURE 4.6 The layout of the building and the buffers that surround it.

4.5 HEIGHT ESTIMATIONS AND QUALITY EVALUATION

The operation for feature height estimation entails creating a variety of products, including DSM, DTM, and building heights. The quality of each of these outputs is evaluated using accuracy parameters. The quality of the DTM is determined by measuring the heights of sample points at various locations throughout the study area. The accuracy of various height estimating systems is compared to manually measured building heights.

4.5.1 DSM Quality Evaluation

The RMSE values are calculated for 42 precise GCPs measured using DGPS over the study area. The RMSE values derived for elevations of DEMs from various sources are listed in Table 4.3. Other elevation sources were found to be less accurate than the height acquired from SRTM and ALSO PRISM. Among all publicly available DEMs, the ASTER data have the most inaccuracies in the region. The RMSE of the DEM derived from Cartosat-1 stereo was 9.85, the highest of all surface models.

TABLE 4.3
Accuracy Metrics Derived from 42 GCPs in the Study Area

Error	Cartosat-30	ASTER	SRTM	ALOS PALSAR	ALOS PRISM	Cartosat-2.5
RMSE	5.26	7.24	4.39	4.57	4.67	9.85

In contrast to previous DEMs that provide geoidal height, Cartosat-30 and ALOS PALSAR provide ellipsoidal height. A constant value of 86.5 m, the average value obtained from DGPS readings, is utilized as an additional factor throughout the study region to convert from ellipsoidal to geoidal elevation. The elevation is provided as a pixel value every 30 m in the publicly available DEMs.

4.5.2 DTM Quality Evaluation

The DTM produced from DSM generated with Cartosat-1 is compared using several approaches. Because all other publicly available DEMs are expected to creates DTM with similar characteristics, they are not included in the quality assessment. Table 4.4 shows the mean difference values of pixel values under 190 building polygons produced using various approaches. The greatest height estimations of the building features are found when elevation is sampled at a 100-m grid. In addition, the observation demonstrates that MSD produces smoother DTM than any other approach.

4.5.3 Building Height Values

Essentially, this chapter uses two DSM approaches to demonstrate building height estimation. The first approach involves subtracting DTM from DSM, while the second employs polygonal buffers around the buildings. The measured heights of buildings are compared to the estimated height to determine the model output's estimation competence. Appendix 4.1 (https://docs.google.com/document/d/1q8XHFBgP3pTpfDl-JUhtt_PAsusiFSvY/edit?usp=sharing&ouid=109926032716705849211&rtpof=true&sd=true)

TABLE 4.4
The Difference in Mean Values of DSM and DTM under 190 Buildings

	Road Buffer	MSD	Slope Filter	Grid 100 m	IDW
Mean difference	1.65	3.73	3.42	7.57	8.14

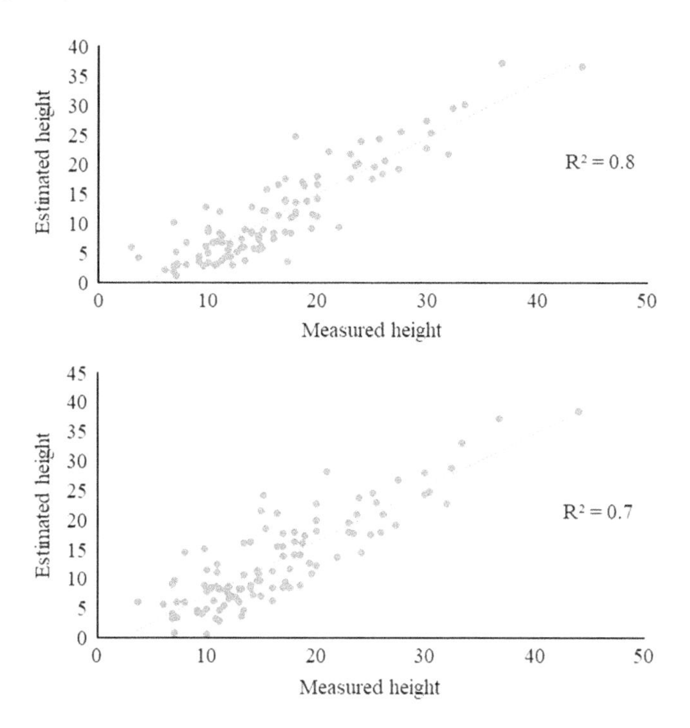

FIGURE 4.7 Estimated and measured building heights using building buffers (top) and grid-based approaches (bottom).

contains information on 190 different building heights. For the current sensor (Cartosat-1) capabilities, an analysis of the building heights produced from the model shows that about 50% of the building heights are estimated within the stipulated error of 10 m. Grid-based DTM generation, followed by an interpolated surface, was the most efficient among the several DTM-generating methods for height estimate.

For a large geographical area, the heights created by the building buffer approach proved to be the most suitable and computationally efficient. The relationship between measured and estimated heights utilizing a grid-based and building buffer technique is shown in Figure 4.7. It implies that the model successfully predicted building heights in the error range of 10–40 m.

4.6 FUTURE SCOPE OF HEIGHT ESTIMATIONS

It is possible that in the future, a similar analysis of the height estimations will be performed on higher resolution optical stereo images, which will allow for coverage of the majority of buildings in the study region. More efficient approaches can be used to produce the surface model. Instead of stereo satellite images, multidate monoscope images can be used for surface model development. The sensor's capacity to capture surface elements such as trees and buildings on the flat and hilly surface conditions must be evaluated and there is a scope for future research. The following chapter will guide in understanding the ability of these sensors to represent the features in three dimensions.

REFERENCES

Asal, F. F. (2019). Comparative analysis of the digital terrain models extracted from airborne LiDAR point clouds using different filtering approaches in residential landscapes. *Advances in Remote Sensing*, 8(02), 51. doi:10.4236/ars.2019.82004

Arefi, H., d'Angelo, P. , Mayer, H . & Reinartz, P. (2011). Iterative approach for efficient digital terrain model production from CARTOSAT-1 stereo images. *Journal of Applied Remote Sensing*, 5(1), 05352710.1117/1.3595265.

Beumier, C., & Idrissa, M. (2015). Deriving a DTM from a DSM by uniform regions and context. *EARSeL eProceedings*, 14(1), 16. doi:10.12760/01-2015-1-02

Gröger, G. H., Kolbe, T., & Czerwinski, A. (Eds.). (2006). Candidate OpenGIS® CityGML implementation specification. *Open Geospatial Consortium.* https://www.ogc.org/

Misra, P., Avtar, R., & Takeuchi, W. (2018). Comparison of digital building height models extracted from AW3D, TanDEM-X, ASTER, and SRTM digital surface models over Yangon City. *Remote Sensing*, 10(12). doi:10.3390/rs10122008

Mousa, A. K., Helmholz, P., & Belton, D. (2017). New DTM extraction approach from airborne images derived DSM. *International Archives of the Photogrammetry, Remote Sensing & Spatial Information Sciences*, 42, 75–82.

Özcan, A. H., Ünsalan, C., & Reinartz, P. (2018). Ground filtering and DTM generation from DSM data using probabilistic voting and segmentation. *International Journal of Remote Sensing*, 39(9), 2860–2883. doi:10.1080/01431161.2018.1434327

Panagiotakis, E., Chrysoulakis, N., Charalampopoulou, V., & Poursanidis, D. (2018). Validation of Pleiades Tri-Stereo DSM in urban areas. *ISPRS International Journal of Geo-Information*, 7(3), 118. doi:10.3390/ijgi7030118

Ranagalage, M., Estoque, R. C., Handayani, H. H., Zhang, X., Morimoto, T., Tadono, T., & Murayama, Y. (2018). Relation between urban volume and land surface temperature: A comparative study of planned and traditional cities in Japan. *Sustainability*, 10(7), 2366.

Sharma, S. A., Agrawal, R., & Jayaprasad, P. (2016). Development of '3D city models' using IRS satellite data. *Journal of the Indian Society of Remote Sensing*, 44(2), 187–196. doi:10.1007/s12524-015-0478-9

Sithole, G., & Vosselman, G. (2004). Experimental comparison of filter algorithms for bare-Earth extraction from airborne laser scanning point clouds. *ISPRS Journal of Photogrammetry and Remote Sensing*, 59(1–2), 85–101. doi:10.1016/j.isprsjprs.2004.05.004

Tsanis, I. K., Seiradakis, K. D., Daliakopoulos, I. N., Grillakis, M. G., & Koutroulis, A. G. (2014). Assessment of GeoEye-1 stereo-pair-generated DEM in flood mapping of an ungauged basin. *Journal of Hydroinformatics*, 16(1), 1–18. doi:10.2166/hydro.2013.197

Vosselman, G. (2000). Slope based filtering of laser altimetry data. *International Archives of Photogrammetry and Remote Sensing*, 33, 935–942.

Wurm, M., d'Angelo, P., Reinartz, P., & Taubenböck, H. (2014). Investigating the applicability of Cartosat-1 DEMs and topographic maps to localize large-area urban mass concentrations. *IEEE Journal of Selected Topics in Applied Earth Observations and Remote Sensing*, 7(10), 4138–4152. doi:10.1109/JSTARS.2014.2346655

Zhang, Y., Zhang, Y., Zhang, Y., & Li, X. (2016). Automatic extraction of DTM from low resolution DSM by two-steps semi-global filtering. *ISPRS Annals of Photogrammetry, Remote Sensing and Spatial Information Science*, 3(3), 249–255. doi:10.5194/isprs-annals-III-3-249-2016

5 3D Feature Mapping

5.1 3D MAPPING FROM GEOSPATIAL DATA

Geospatial technology and associated digital information play an important role in today's urban scenario. The dense urban setups in several developing countries are causing severe infrastructural problems. Rapid organic development and bad planning are important contributors to our cities' sorry state of affairs (Bharath et al., 2018; Ramachandra et al., 2017). The governing authorities are hampered by a scarcity of accurate statistics to make educated decisions. In addition to many other technological solutions, geospatial technology has grown in appeal across a wide range of stakeholders. The primary reason for this is a faster method for obtaining large size datasets and quick updating capabilities. The application of 3D mapping using remote sensing technology transformed the way data were collected and analysed. The capacity to map buildings, trees, towers, bridges, and a variety of other structures has provided much-needed dependable datasets for a variety of development tasks. Significant research has gone into remote sensing data collection platforms (satellite, airborne, and terrestrial) or sensors (multispectral, Lidar, radar, etc.). Furthermore, tools for automatically converting raw data into usable maps are fast evolving.

Megacities in emerging countries are expanding in the third dimension, in addition to horizontal sprawl. The large expansion of built-up area and building development are the key contributors to urban growth in land-use analysis. Understanding and modelling these urban development phenomena can help with improved planning and giving long-term solutions. Several parties have expressed interest in portraying a building on a map as a three-dimensional object. The inclusion of a third dimension to digital mapping tends to complicate feature representation and relationships between other spatial elements. The complex urban system is made up of many relatively tiny spatial objects such as buildings, parks, roads, street lights, and so on. As a result of its ability to leverage topographical correlations, spatial technology to represent urban systems is prevalent. Man-made features, as opposed to natural earth objects such as trees, rivers, and hills, are regular in terms of structure, shape, and distribution. This aspect of urban features makes it easier for technology developers to express urban characteristics as spatial objects with precise geometry.

DOI: 10.1201/9781003288046-5

The following sections cover many features of 3D building mapping (LOD-1), a brief history of 3D map production from its beginning to the present, current data standards, and interoperability difficulties. Various tools and tactics for building 3D models are discussed because visualization is important to the success of 3D map-based applications. This chapter describes and demonstrates the use of geospatial technology to produce 3D building models and its benefits in the real world.

5.2 HISTORY OF 3D MAPPING

During the British Ordnance Survey and the Great Trigonometrical Survey of India in the seventeenth century, the concept of scaling maps, defining datum, and recording elevations were pioneered (Garfield, 2013). Despite important advances in photogrammetry by French scientist Aimé Laussedat (called the 'Father of Photogrammetry') in 1849, the first stereo photograph was taken by Nadar during Emperor Napoleon's reign in the 1870s. However, photogrammetry was initially used for reconnaissance rather than measurement. In 1885, George Eastman, the founder of the Kodak Corporation and the inventor of the floating mark, performed the first stereoscopic measurements. Collier (2002) mentions the early development of photogrammetry and the measurement of elevation information in the form of contours in his handbook *History of Photogrammetry*. The period of analogue photogrammetry began in the first quarter of the twentieth century, when various inventions ushered in substantial advancements in photogrammetric surveying. Despite the fact that the essential principles of analytical photogrammetry were created between 1900 and 1950, analytical photogrammetry was not widely used until the 1950s. During the last quarter of the twentieth century, digital photogrammetry arose in tandem with the advancement of computers.

The history of current mapping concepts can be traced back to Roger Tomlinson's Canada Geographic Information System (CGIS) in the late 1950s (Laura, 2018; Cracknell, 2018). During the 1970s, the marriage of analytical photogrammetry and geographic information systems (GIS) took place, and the instrument developed was known as STARS (Simultaneous Triangulation and Resection Software). However, until the 1990s, computerized maps were mostly two-dimensional in depiction. Early USGS maps, known as the 7.5-minute series, featured elevation information as contours to illustrate topographical differences and served as the foundation for modern 3D mapping. SYMVU, a 3D perspective view of SYMAP output to visualize population density of the United States during 1970 developed by Harvard Lab for Computer Graphics and Spatial Analysis, was the first 3D visualization (ESRI).

American corporations created the first 3D maps, one of which was Silicon Graphics in the early 1990s (Goodchild, 2018). It should be noted that the establishment of the Global Positioning System (GPS), which has a constellation called Navigation System with Timing and Ranging (NAVSTAR) satellites launched in 1978 and was completely operational by 1993, enabled the worldwide measurement of co-ordinates in 3D space. Overall, technological advancements in space-based remote sensing techniques, particularly stereo imaging, proved to be effective methods for obtaining digital 3D surface information over the last quarter of the twentieth century (Madry, 2017).

The 1980s saw the development of computer vision technology for 3D surface reconstruction, but the last decade of the twentieth century saw the widespread adoption of digital photogrammetry. Stereo imaging satellites such as IKONOS, Cartosat, and PLEIADES were launched, further popularizing digital 3D data creation. The ISO 19115 standard was established in 1997, and it includes standards for 3D data visualization such as X3D and VRML, among others. In the early twenty-first century, the Open Geospatial Consortium (OGC) developed the 3D information management working group, which included service providers from CAD/GIS software developers and European administrative and government institutions. With the advent of Google Earth in 2005, which was previously known as Keyhole EarthViewer by its manufacturer business Keyhole Inc., millions of users were able to experience 3D maps for the first time. Digital photogrammetry, computer vision, and aerial photos taken with drones have revolutionized 3D mapping technology during the last two decades.

In the year 2009, Oracle released the 3D spatial engine (Ravada et al., 2009), an exclusive database for geospatial databases. PostGree added a PostGIS v2.0 extension that allows 3D datasets at the same time. Later, a standard relational database called 3D City DB was built to represent virtual 3D city models, with the ability to display CityGML and convert between GTiff, KML, and COLLADA formats. CityJSON (Ledoux, 2018) was recently established with the goal of resolving the complications in CityGML, and has receiveda lot of attention from the developer community. Because of the benefits indicated in the article, the OGC is working to make it a standard data representation and exchange format (Arroyo Ohori, 2020). According to the article after 2008, the development of several viewers (FZK Viewer, LandXplorer, and GML Viewer) followed the CityGML data format.

5.3 DATA STANDARDS AND INTEROPERABILITY

The geospatial ecosystem is rapidly moving to a 3D environment in various application domains. The increasing availability of high-resolution images, computing capabilities of graphical processing units (GPU), and the development of open-source software applications have sparked interest in 3D models among various stakeholders. Since the increasing adoption of digital maps, attempts have been made to improve the representation of earth features in a more accurate and user-friendly manner. Over the last two decades, both stand-alone software tools and web browser-based applications have made significant advances in 3D data representation. Developers' attempts to incorporate geometrical, pictorial, and semantic components in the 3D model data standard for multiple applications have been documented in numerous scientific articles. The data standards were unavoidable because the full lifecycle of 3D models passes through various stakeholders, from data creators to decision makers. Furthermore, the data preparation stage necessitates collaboration among several human groups in order to produce useful 3D models.

The OGC accepted CityGML as the international standard for 3D city model representation and interchange in 2008. The CityGML is intended to spatially represent various relevant urban objects (buildings, bridges, furniture, and so on) so that the model can communicate aesthetically, geometrically, and semantically. A building's representation on a map can take numerous forms, depending on the level of information (LOD). Figure 5.1 shows a graphical representation of various levels of detail. The geometry and semantics of the characteristics are combined in the LOD-based notion of building representation, with increasing details from 0 to 4. It is obvious that the LOD-3 features pertaining to building interiors are less important in broad area mapping. The exact geometry of individual building roof structures depicted in LOD-2 is required for many niche applications, which is outside the focus of this book. In 3D mapping of buildings utilizing geospatial technologies, which are represented by LOD-1, the height and horizontal distribution of building footprints are significant. The LOD-1 representation is practical because of its simplicity and smoother processing on a computer tool.

Nonetheless, the LOD-based representation of 3D city models, particularly architectural characteristics, faces a number of obstacles, including ambiguity created by data collection techniques (Biljecki et al., 2016),

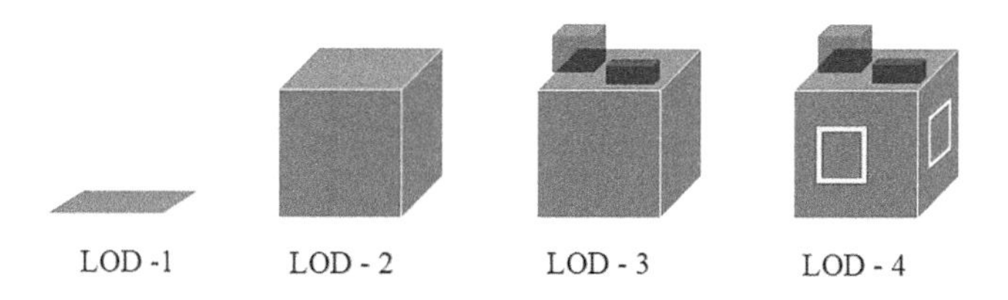

FIGURE 5.1 According to OGC standards, the levels of detail for buildings.

stakeholder misunderstanding, and a wide range of application use case requirements. The data creator must consider numerous real-world elements in model development such as dimensions, texture, list of attributes, spatio-semantic coherence, and feature complexity while creating the models in various LODs (Biljecki et al., 2014). Also, the authors' main opinion is that it is impossible to set requirements for individual LOD models because different applications require different criteria. Labetski et al. (2018) describe the importance of metadata in CityGML and propose a framework for storing it in the ADE. Due to the limited hierarchy of representation, 3D city models created using CityGML have some restrictions in depicting non-building characteristics such as vegetation, water bodies, land use, and so on (Arroyo Ohori, 2020). Due to limited support from open-source tools and obstacles in data generation, such as expensive costs and labour-intensive tasks, the widespread adoption of 3D city models utilizing CityGML has slowed (Vitalis et al., 2020). Coors et al. (2020) discussed several parts of the CityGML file quality check based on individual needs or application use cases. Malinverni et al. (2020) propose a graph-based database for achieving efficient connectivity between GIS and BIM models based on the capabilities of 3D building models represented using CityGML to execute spatial analysis on modelled terrain (Malinverni et al., 2020). In their study, Yao et al. (2018) discuss the necessity for a special purpose database to handle 3D city models across many platforms. Scianna (2013) suggested a three-dimensional model for spatial objects that may be visualized using web browsers. Van Den et al. (2013) address high-level features of establishing a national standard for 3D topographic data representation using CityGML for the Netherlands in their study. Goetz (2013) created a method for developing extremely detailed 3D CityGML models using OpenStreetMap datasets.

The OGC CityGML models provide a viable standard for meeting organizations' geographic data infrastructure needs. Because it is interoperable, data may be disseminated across multiple platforms with few problems. The growth of internet technology and the migration of everything to cloud-based platforms have made data interoperability a must-have. Internet-based solutions offer scalability in terms of data, users, processing, and various other factors. The problems with CityGML are as follows:

- Models are complex
- Database connections are inefficient
- Data security and integrity concerns
- Data transfer across domains is inefficient

The open-source database 3DCityDB was created at the University of Bonn in 2003 and is now utilized in a variety of applications (Yao et al., 2018). The 3DCityDB was widely accepted in providing datasets over web applications since it supports the CityGML standards (Beil and Kolbe, 2017). The application domain extension (ADE) mechanism is available to make appropriate links to external applications, and the authors specify 44 ADEs for CityGML by that time (Biljecki et al., 2018). The restrictions of dealing with CityGML on web platforms, which were once a source of concern, have been effectively addressed by CityJSON. It allows you to connect to external databases without changing your 3D models.

CityJSON (Ledoux, 2018; Ledoux et al., 2019) was recently designed to address the complexity of CityGML and has received a lot of attention from the developer community. Because of the benefits indicated in the article, the OGC is working to make it a standard data representation and exchange format (Arroyo Ohori, 2020). CityJSON is built on JavaScript (JSON) and is designed to be versatile in terms of representing and transferring data across diverse software and web applications. For example, Ohori et al. (2015) created a mapping between multiple LODs to produce 4D model primitives. The 4D model is made up of (x,y,z,l), with l representing a point on the LOD axis. In the conclusion section, the authors argue that 3D city models are still in their infancy, implying that more effort is required to build practical applications. Breunig et al. (2020) present an overview of recent breakthroughs in 3D/4D geographic data management as well as future directions.

5.4 DATA SOURCES FOR 3D MAPPING

When considering 3D mapping of buildings up to LOD-1, two major details are needed: firstly, a polygonal outline of the building and secondly height value. These details can be practically obtained from physical measurements and remote sensing techniques. Among remote sensing methods, Lidar point clouds are utilized for precise details due to their high accuracy measurements. Stereo satellite images and aerial images are other ways to get accurate height information. However, its largely hard to obtain both polygonal outline shape of building roofs and heights values using automated methods. Large-scale industrial applications mainly involve using Lidar and manual digitization or data processing methods to obtain data necessary for 3D mapping buildings. Drone images are low in cost and a reliable way of information generation in this regard. However, increasing satellite image resolution, stereo or tristereo imaging, and multiple overlapping scenes at very high resolution opens up plenty of scope for generating height maps as digital

TABLE 5.1
Prominent Online Resources for 3D Map Ready Datasets

Name	Details	Format
Open City Model	All buildings of USA LOD1	GML, JSON
NYC 3D building model	Buildings of New York City, Up to LOD2	CityGML, ESRI Multipatch, DGN
Random3D City	Synthetic CityGML data and procedural modelling engine	CityGML
OpenStreetMap	Across multiple cities, commercial data provider	Shapefile, JSON
VISICOM	Cross world, on request, commercial	Shapefile, JSON
TUDelft	Open datasets repository, multiple cities, LOD1 or LOD2	CityJSON, Shapefile

surface models. A list of stereo imaging satellite sources is given in Table 4.1 (Chapter 4), and few online ready to use 3D map ready datasets are provided in Table 5.1.

5.5 SOFTWARE TOOLS FOR 3D MAPPING

Brooks and Whalley (2008) created a hybrid software tool that can effortlessly switch between 2D and 3D views of a scene. Glander and Döllner (2009) developed an abstract representation of 3D city models, including buildings, infrastructure networks, and land covering zones. Wu et al. (2010) created an interactive 3D representation to understand the urban development process better. The technological components of 3D visualization and computer graphics techniques relevant to the depiction of surfaces with spatial features are discussed by Lorenz and Döllner (2010). Mao et al. (2011) proposed a data structure paradigm that would allow dynamic zooming of 3D city models. Trubka et al. (2016) created Envision Scenario Planner (ESP), a web-based 3D visualization tool that facilitates assessment system modelling. Hildebrandt (2014) proposed software reference architectural approaches for service-oriented 3D geovisualization systems. Virtanen et al. (2015) created collaborative software that links 3D models and attributes for efficient management. Zhou et al. (2017) examined features of parallelism principles in LOD-based 3D model display on GPU-based computing systems. Wang et al. (2018) showed the technology of using Microsoft Hololens to display 3D geographic

TABLE 5.2
Sophisticated Software Tools for 3D Urban Datasets

Tools	Functionalities
cjio (commond line)	Query, statistics, metadata, compression, subset, decompression
citygmltools	Conversion between CityGML and CityJSON
Azul	View 3D models in macOS
Val3dity	Validations of 3D model primitives
3dfier	Creation of 3D city models in CityJSON from 2D datasets
QGIS	This is a QGIS plugin that visualizes DEM and vector data in 3D on web browsers and in QGIS window
3DCityDB	Translating CityGML data to a relational database
VI-Suit Blender	Solar irradiance mapping and building modelling, analysis of geospatial data
Blender CityJSON plugin	Creating an immersive gaming experience with 3D city models

information through a tool named Holo3DGIS. Neuville et al. (2018) detail the formalization of 3D geovisualization by increasing the efficiency of visualization and recognition of specific objects. Yu et al. (2012) created a hybrid model that can render 2D and 3D objects in a single window and includes features such as querying, 2D analysis in a 3D environment, and more. Ruzinoor et al. (2012) conducted a comprehensive study of geospatial data visualization 3D visualization technologies. For efficient 3D modelling, Guo et al. (2016) proposed an event-driven spatiotemporal database approach. Liang et al. (2017) devised a three-step strategy for repairing bespoke alterations in existing city models with minimal human interaction. Zhang et al. (2011) describe GeoScope, as a comprehensive 3D visualization tool in a GIS. Zhang et al. (2014) created a web application that allows users to view massive 3D building models (Table 5.2).

5.6 EXPERIMENTS

Various approaches for creating LOD-1 3D building datasets are discussed in this section. The approaches used in the literature can be classified into two groups. Polygonal outlines and height information are among the building primitives. Figure 5.2 depicts the entire data preparation and visualization process. Manual digitization from cadastral maps, remote sensing images, and other technologies are used to create the building outlines.

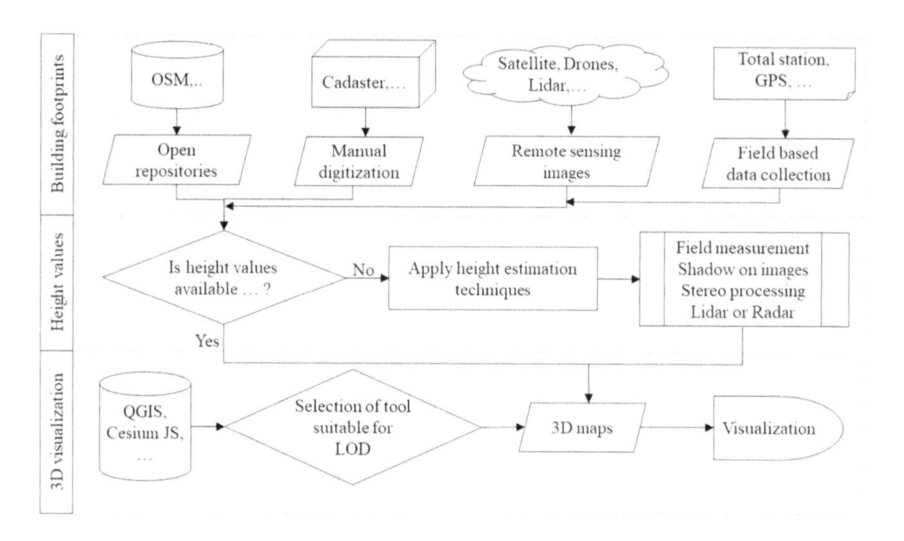

FIGURE 5.2 The process of 3D data preparation and visualization.

In some cases, field-based surveys and existing digital maps are used. Building height data can be gathered from a variety of sources, including stereo satellite images, synthetic aperture radar (SAR), drone-based images, and Lidar imaging techniques (Ramiya et al., 2017). Satellite images, on the other hand, cover a larger region, whereas aerial surveys are more accurate and contain greater details.

Cao et al. (2017) used aerial Lidar point clouds data to illustrate roof reconstruction methods. Bagheri et al. (2019) describe how to improve accuracy using a fusion of SAR and imaging datasets. Using TerraSAR-X images, Wu et al. (2019) demonstrated 3D visualization of high-rise buildings. Guo et al. (2019) used interferometric SAR images to recreate 3D models by estimating the height of high-rise buildings. Sharafzadeh et al. (2018) sought to create 3D city models with varied attributes and height information using SAR datasets. For efficient 3D reconstruction, Yastikli and Cetin (2021) used the automatic classification of roof points from the Lidar dataset. Guo et al. (2021) reconstructed urban buildings using high-resolution tomographic synthetic aperture radar (TomoSAR). Satellite and aerial images can be used to construct building outlines, as demonstrated. To visualize 3D block models, manually digitized vector features and matching height values produced from a digital surface model are used.

Because of the realistic experience, the representation of 3D buildings on a map setting grabs the audience's interest right away. Since the inception of digital maps, technology developers have worked to create intuitive visualizations as simple as possible. The majority of premier GIS software packages provide the ability to work with 3D city models

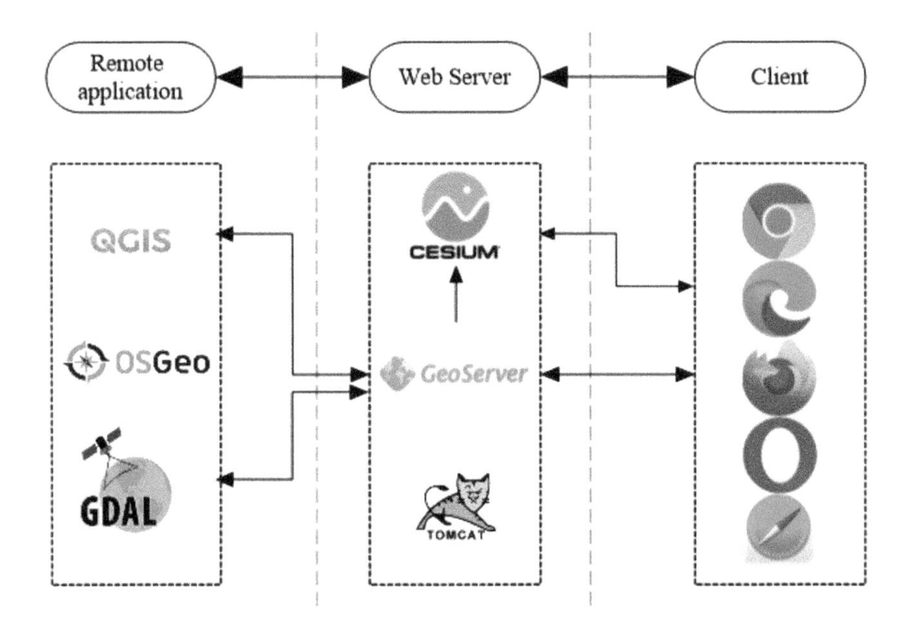

FIGURE 5.3 Web architecture for visualization of 3D geospatial data.

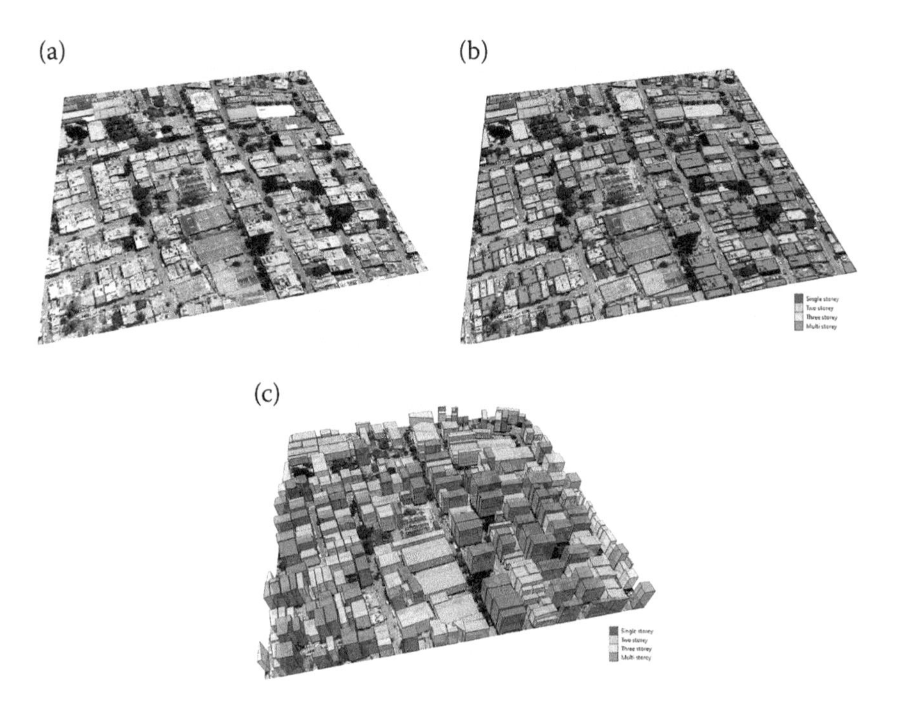

FIGURE 5.4 (a) Orthophoto showing buildings; (b) building footprints overlaid on an orthophoto; (c) 3D map visualization of buildings.

that include prominent building features. On the other hand, open-source software such as CesiumJS, Keplergl, Blender, and QGIS provides the required tools for creating effective 3D visualizations.

The QGIS plugin Qgis2threejs and CesiumJS are used to demonstrate 3D building mapping. Along with DEM, the QGIS plugin allows to visualize 3D features. CesiumJS is a free and open-source JavaScript toolkit for creating 3D globes and maps. CesiumJS may be used to build interactive 3D online apps that share dynamic geospatial data. Figure 5.3 depicts the architecture utilized to create the application. Figure 5.4 shows the 3D map visualizations made using QGIS environment.

REFERENCES

3DCityDB, https://www.3dcitydb.org/3dcitydb/

3dfier, https://github.com/tudelft3d/3dfier

Arroyo Ohori, K. (2020). Azul: A fast and efficient 3D city model viewer for macOS. *Transactions in GIS*, 24(5), 1165–1184. doi:10.1111/tgis.12673

Azul, https://github.com/tudelft3d/azul

Bagheri, H., Schmitt, M., & Zhu, X. (2019). Fusion of multi-sensor-derived heights and OSM-derived building footprints for urban 3D reconstruction. *ISPRS International Journal of Geo-Information*, 8(4), 193.

Beil, C., & Kolbe, T. H. (2017). CityGML and the streets of New York: A proposal for detailed street space modelling. In *Proceedings of the 12th International 3D GeoInfo Conference 2017*, pp. 9–16. doi:10.5194/isprs-annals-IV-4-W5-9-2017

Bharath, H. A., Chandan, M. C., Vinay, S., & Ramachandra, T. V. (2018). Modelling urban dynamics in rapidly urbanising Indian cities. *The Egyptian Journal of Remote Sensing and Space Science*, 21(3), 201–210. doi:10.1016/j.ejrs.2017.08.002

Biljecki, F., Kumar, K., & Nagel, C. (2018). CityGML application domain extension (ADE): Overview of developments. *Open Geospatial Data, Software and Standards*, 3(1), 1–17. doi:10.1186/s40965-018-0055-6

Biljecki, F., Ledoux, H., & Stoter, J. (2016). An improved LOD specification for 3D building models. *Computers, Environment and Urban Systems*, 59, 25–37. doi:10.1016/j.compenvurbsys.2016.04.005

Biljecki, F., Ledoux, H., Stoter, J., & Zhao, J. (2014). Formalisation of the level of detail in 3D city modelling. *Computers, Environment and Urban Systems*, 48, 1–15. doi:10.1016/j.compenvurbsys.2014.05.004

Blender CityJSON plugin, https://github.com/cityjson/Up3date

Brooks, S., & Whalley, J. L. (2008). Multilayer hybrid visualizations to support 3D GIS. *Computers, Environment and Urban Systems*, 32(4), 278–292.

Breunig, M, Bradley, P. E., Jahn, M., Kuper, P., Mazroob, N., Rösch, N., Al-Doori, M., Stefanakis, Emmanuel, & Jadidi, M. (2020). Geospatial data management research: Progress and future directions. *ISPRS International Journal of Geo-Information*, 9(9510.3390/ijgi9020095.

Cao, R., Zhang, Y., Liu, X., & Zhao, Z. (2017). 3D building roof reconstruction from airborne LiDAR point clouds: A framework based on a spatial database. *International Journal of Geographical Information Science*, 31(7), 1359–1380.

citygmltools, https://github.com/citygml4j/cityg ml-tools

cjio, https://github.com/tudelft3d/cjio

Collier, P. (2002). The impact on topographic mapping of developments in land and air survey: 1900–1939. *Cartography and Geographic Information Science*, 29(3), 155–174. doi:10.1559/152304002782008440

Coors, V., Betz, M., & Duminil, E. (2020). A concept of quality management of 3D city models supporting application-specific requirements. *PFG–Journal of Photogrammetry, Remote Sensing and Geoinformation Science*, 88(1), 3–14. doi:10.1007/s41064-020-00094-0

Cracknell, A. P. (2018). The development of remote sensing in the last 40 years. *International Journal of Remote Sensing*, 39(23), 8387–8427. doi:10.1080/01431161.2018.1550919

Garfield, S. (2013). *On the Map: A Mind-Expanding Exploration of the Way the World Looks*. The Wilson Quarterly.

Glander, T., & Döllner, J. (2009). Abstract representations for interactive visualization of virtual 3D city models. *Computers, Environment and Urban Systems*, 33(5), 375–387.

Goetz, M. (2013). Towards generating highly detailed 3D CityGML models from OpenStreetMap. *International Journal of Geographical Information Science*, 27(5), 845–865. doi:10.1080/13658816.2012.721552

Goodchild, M. F. (2018). Reimagining the history of GIS. *Annals of GIS*, 24(1), 1–8. doi:10.1080/19475683.2018.1424737

Guo, H., Li, X., Wang, W., Lv, Z., Wu, C., & Xu, W. (2016). An event-driven dynamic updating method for 3D geo-databases. *Geo-spatial Information Science*, 19(2), 140–147.

Guo, R., Wang, F., Zang, B., Jing, G., & Xing, M. (2019). High-rise building 3D reconstruction with the wrapped interferometric phase. *Sensors*, 19(6), 1439. doi:10.3390/s19061439

Guo, Z., Liu, H., Pang, L., Fang, L., & Dou, W. (2021). DBSCAN-based point cloud extraction for tomographic synthetic aperture radar (TomoSAR) three-dimensional (3D) building reconstruction. *International Journal of Remote Sensing*, 42(6), 2327–2349.

Hildebrandt, D. (2014). A software reference architecture for service-oriented 3D geovisualization systems. *ISPRS International Journal of Geo-Information*, 3(4), 1445–1490.

Labetski, A., Kumar, K., Ledoux, H., & Stoter, J. (2018). A metadata ADE for CityGML. *Open Geospatial Data, Software and Standards*, 3(1), 1–16. doi:10.1186/s40965-018-0057-4

Laura, T. (2018, March 21). An Overview of GIS History. *Geospatial World*. https://www.geospatialworld.net/blogs/overview-of-gis-history/

Ledoux, H. (2018). val3dity: Validation of 3D GIS primitives according to the international standards. *Open Geospatial Data, Software and Standards*, 3(1), 1–12. doi:10.1186/s40965-018-0043-x

Ledoux, H., Ohori, K. A., Kumar, K., Dukai, B., Labetski, A., & Vitalis, S. (2019). CityJSON: A compact and easy-to-use encoding of the CityGML data model. *Open Geospatial Data, Software and Standards*, 4(1), 1–12. doi:10.1186/s40965-019-0064-0

Lorenz, H., & Döllner, J. (2010). 3D feature surface properties and their application in geovisualization. *Computers, Environment and Urban Systems*, 34(6), 476–483.

Liang, J., Shen, S., Gong, J., Liu, J., & Zhang, J. (2016). Embedding user-generated content into oblique airborne photogrammetry-based 3D city model. *International Journal of Geographical Information Science*, 31, 1–1610.1080/13658816.2016.1180389.

Madry, S. (2017). Introduction and history of space remote sensing. *Handbook of Satellite Applications*, 823. doi:10.1007/978-3-319-23386-4_37

Malinverni, E. S., Naticchia, B., Garcia, J. L. L., Gorreja, A., Uriarte, J. L., & Di Stefano, F. (2020). A semantic graph database for the interoperability of 3D GIS data. *Applied Geomatics*, 1–14. doi:10.1007/s12518-020-00334-3

Mao, B., Ban, Y., & Harrie, L. (2011). A multiple representation data structure for dynamic visualisation of generalised 3D city models. *ISPRS Journal of Photogrammetry and Remote Sensing*, 66, 198–20810.1016/j.isprsjprs.2010.08.001.

Neuville, R., Pouliot, J., Poux, F., De Rudder, L., & Billen, R. (2018). A formalized 3D geovisualization illustrated to selectivity purpose of virtual 3D city model. *ISPRS International Journal of Geo-Information*, 7(5), 194.

NYC 3D building model, https://www1.nyc.gov/site/doitt/initiatives/3d-building.page

Open City Model, https://github.com/opencitymodel/opencitymodel

Ohori, K., Ledoux, H., Biljecki, F., & Stoter, J. (2015). Modeling a 3D city model and its levels of detail as a true 4D model. *ISPRS International Journal of Geo-Information*, 4, 1055–107510.3390/ijgi4031055.

OpenStreetMap 3D, https://3dbuildings.com/data/

OpenStreetMap, https://osmbuildings.org/data/

QGIS, https://github.com/minorua/Qgis2threejs

Ramachandra, T. V., Bajpai, V., Kulkarni, G., Aithal, B. H., & Han, S. S. (2017). Economic disparity and CO2 emissions: The domestic energy sector in Greater Bangalore, India. *Renewable and Sustainable Energy Reviews*, 67, 1331–1344. doi:10.1016/j.rser.2016.09.038

Ravada, S., Kazar, B. M., & Kothuri, R. (2009). Query processing in 3D spatial databases: Experiences with oracle spatial 11g. In *3D Geo-Information Science*. (pp. 153-173).Springer, Berlin: Heidelberg.

Ramiya, A. M., Nidamanuri, R. R., & Krishnan, R. (2017). Segmentation based building detection approach from LiDAR point cloud. *The Egyptian Journal of Remote Sensing and Space Science*, 20(1), 71–77. doi:10.1016/j.ejrs.2016.04.001

Random3Dcity, https://filipbiljecki.com/code/Random3Dcity.html

Ruzinoor, C. M., Shariff, A. R. M., Pradhan, B., Todzi Ahmad, M., & Rahim, M. S. M. (2012). A review on 3D terrain visualization of GIS data: Techniques and software. *Geo-Spatial Information Science*, 15(2), 105–115.

Scianna, A. (2013). Building 3D GIS data models using open source software. *Applied Geomatics*, 5(2), 119–132. doi:10.1007/s12518-013-0099-3

Sharafzadeh, A., Esmaeily, A., & Dehghani, M. (2018). 3D modelling of urban area using synthetic aperture radar (SAR). *Journal of the Indian Society of Remote Sensing*, 46(11), 1785–1793.

Trubka, R., Glackin, S., Lade, O., & Pettit, C. (2016). A web-based 3D visualisation and assessment system for urban precinct scenario modelling. *ISPRS Journal of Photogrammetry and Remote Sensing*, 117, 175–186.

TUDelft, https://3d.bk.tudelft.nl/opendata/opencities/

Val3dity, https://github.com/tudelft3d/val3dity

Van den Brink, L., Stoter, J., & Zlatanova, S. (2013). Establishing a national standard for 3D topographic data compliant to CityGML. *International Journal of Geographical Information Science*, 27(1), 92–113. doi:10.1080/13658816.2012.667105

VISICOM, https://visicomdata.com

VI-Suite Blender, https://blogs.brighton.ac.uk/visuite/

Vitalis, S., Arroyo Ohori, K., & Stoter, J. (2020). CityJSON in QGIS: Development of an open-source plugin. *Transactions in GIS*, 24(5), 1147–1164. doi:10.1111/tgis.12657

Virtanen, J.-P., Hyyppä, H., Kämäräinen, A., Hollström, T., Vastaranta, M., & Hyyppä, J. (2015). Intelligent open data 3D maps in a collaborative virtual world. *ISPRS International Journal of Geo-Information*, 4, 837–857. 10.3390/ijgi4020837.

Wang, W., Wu, X., Chen, G., & Chen, Z. (2018). Holo3DGIS: Leveraging Microsoft Hololens in 3D geographic information. *ISPRS International Journal of Geo-Information*, 7(2), 60.

Wu, H., He, Z., & Gong, J. (2010). A virtual globe-based 3D visualization and interactive framework for public participation in urban planning processes. *Computers, Environment and Urban Systems*, 34(4), 291–298.

Wu, C., & Hu.Y. (2019). Detection and 3D visualization of deformations for high-rise buildings in Shenzhen, China from high-resolution TerraSAR-X datasets. *Applied Sciences*, 9, 381810.3390/app9183818.

Yao, Z., Nagel, C., Kunde, F., Hudra, G., Willkomm, P., Donaubauer, A., ... & Kolbe, T. H. (2018). 3DCityDB-a 3D geodatabase solution for the management, analysis, and visualization of semantic 3D city models based on CityGML. *Open Geospatial Data, Software and Standards*, 3(1), 1–26. doi:10.1186/s40965-018-0046-7

Yastikli, N., & Cetin, Z. (2021). Classification of raw LiDAR point cloud using point-based methods with spatial features for 3D building reconstruction. *Arabian Journal of Geosciences*, 14(3), 1–14.

Yu, L. J., Sun, D. F., Peng, Z. R., & Zhang, J. (2012). A hybrid system of expanding 2D GIS into 3D space. *Cartography and Geographic Information Science*, 39(3), 140–153.

Zhang, L., Han, C., Zhang, L., Zhang, X., & Li, J. (2014). Web-based visualization of large 3D urban building models. *International Journal of Digital Earth*, 7(1), 53–67.

Zhang, Y., Zhu, Q., Liu, G., Zheng, W., Li, Z., & Du, Z. (2011). GeoScope: Full 3D geospatial information system case study. *Geo-Spatial Information Science*, 14(2), 150–156.

Zhou, D., Jiang, G., Yu, J., Liu, L., & Li, W. (2017). The analysis of task and data characteristic and the collaborative processing method in real-time visualization pipeline of urban 3DGIS. *ISPRS International Journal of Geo-Information*, 6(3), 69.

6 Application Use Cases

6.1 POTENTIAL APPLICATIONS

The urban environment has been strained by the migration of people from rural communities to cities. These events resulted in unrestrained urbanization and deterioration of lifestyle and environment. Several Indian cities are seeing significant growth (Bharath et al., 2017), leading the government to launch a spate of rehabilitation projects to meet people's increasing needs. One of the government's many programs is to increase the use of renewable energy. The implementation of comparable programs necessitates collecting bankable data on urban structures to carry out initiatives and achieve the desired goals. In urban retrofitting and rehabilitation, affordable housing, water supply, reliable data connectivity, digitization, and transparency are significant thrust areas (Ramachandra and Sudhira, 2011). Utility management, transportation facilities, water supply, renewable energy, and slum redevelopment are critical components of area-based development initiatives of smart city projects.

In recent years, dealing with unauthorized construction and demolition difficulties has become a popular pastime. Building structures on encroached land is a long-standing issue that the government, municipalities, and citizens are battling. It has become necessary to identify buildings owned by developers who break the rules for financial advantage by violating zonal and construction standards; this responsibility falls to city governments. The height of the building, the nature of the land use, the percentage of the plot used for construction, and other elements are all significant violations here (Harini, 2017). City models portraying the urban setup are essential to perform rescue, relief, and restoration efforts after natural catastrophes such as floods. The database model heavily relies on pre- and post-disaster data about building type, size, shape, and related information (Dash et al., 2004). With urban volume data or 3D models, environmental phenomena such as urban heat islands and urban land surface temperature research (Ranagalage et al., 2018) would become more accurate.

Traditional field-based survey procedures, as well as manual digitizing processes, are time-consuming and tiresome. In terms of speed, replicability, and model-based approach, algorithm-based machine learning techniques, now widely used in urban feature extraction, have

DOI: 10.1201/9781003288046-6

significant advantages over prior methods. Remotely sensed images typically deal with a lot of information. As a result, machine learning approaches are optimal for categorization and prediction. Machine learning algorithms can handle large amounts of data with a variety of features. The study of machine learning in the field of geospatial technology has become more fascinating due to the ever-increasing number of data, the demand for faster response times, the rising expense of the human workforce, and robust processing capacity. To meet the increased need for technology solutions for detailed urban structure extraction, continuous advances in methodologies are required. This book suggests using machine learning approaches to construct models to extract building information from optical remote sensing imagery automatically. In addition, in prior chapters, stereo satellite methods for generating building height information were addressed. In the following sections, we'll go into a few case studies that call for developing data generated by the models mentioned earlier.

6.2 CASE STUDY #1: URBAN STRUCTURE EXTRACTION – AN INDIAN CONTEXT

Over the millennia, cities have been the lifeblood of human civilization, facilitating important trade-related activity. Industrialization in the late seventeenth-century boosted the growth of cities and urban centres worldwide. More people are migrating to cities due to population expansions, improved lifestyles, and the availability of healthcare facilities. This phenomenon is indeed evident in India. Several megacities sprang up randomly under British colonization worldwide and later in independent India (Bharath et al., 2017; Chandan et al., 2020). People began to migrate from the agrarian system to work in the industrial sector and cities. Cities' fundamental infrastructure development has not been able to keep up with population densification.

Infrastructure development that is well-planned and executed can help the country achieve long-term urban growth and advancement. The amount of people moving to cities necessitates the construction of adequate housing and other infrastructure. Informal settlements cause urban sprawl because of poor populations and unplanned growth (Bharath et al., 2017). According to a Work Bank assessment, an extra 1.2 million square kilometres of land will be added to the urban setup during the next three decades. This has put a lot of strain on the urban ecology, resulting in a lot of land-use changes and the current situation.

Cities are susceptible to climate change events such as sea-level rise, urban heat islands, disasters, and pandemics because of their dense populations. Compared to rural environments, the nature of informal and unplanned settlements in cities is more hazardous. The problem is exacerbated by the fact that many major cities worldwide, including India, are located along the ocean, and river shores add to the problem. Furthermore, due to a lack of horizontal space, the geographical locations of certain cities necessitate vertical growth. Building and road infrastructure in central metropolitan districts and planned periphery developments are key variables for long-term urban development. In this environment, the creation of sustainable cities takes precedence in order to meet the rising population's demands.

'Make cities and human settlements inclusive, safe, resilient, and sustainable', says Sustainable Development Goal 11. This emphasis on urban regions is appropriate, and it is widespread in present Indian city conditions. According to government data, there was a 2,774 town rise from 2001 to 2011. In 2011, a High-Powered Expert Committee (HPEC) suggested to the Ministry of Urban Development that extra investment of $40 billion be made in constructing urban infrastructure over the following two decades (Isher, 2011). This was followed by a slew of reforms and mission-style initiatives, including the Jawaharlal Nehru National Urban Renewal Mission (JNNURM), which was renamed the Atal Mission for Rejuvenation and Urban Transformation (AMRUT) with a broader vision and set of objectives. Changes in law or acts and administrative, structural, and e-governance reforms are among the reforms. The government then launched a series of missions aimed at urban development, including large-scale reconstruction projects. Figure 1.4 depicts the missions announced to transform the urban landscape in Indian cities (MoHUA, 2014). India has taken various measures to achieve the SDGs, including the Ganga Rejuvenation Project, the National Solar Mission, Smart Cities, and the Swachh Bharat Mission (Niti Ayog, 2016). India wants to achieve a 40-gigawatt rooftop solar photovoltaic energy target by 2022 (MNRE, 2014). According to the Ministry of Renewable Energy (MNRE), 6.00 GW will be installed by the end of 2020, for a total capacity of 32 GW. According to the Ministry of Water Resources, rainwater harvesting facilities must be implemented by local governments in urban areas (MOWR, 2013). In addition, the government has undertaken several reconstruction programs in housing, transportation, public spaces, and safety to make Indian cities more sustainable and enhance living standards. Various housing and redevelopment initiatives have been assigned to India's Ministry of Housing and Urban Affairs (MoHUA), which requires up-to-date data to plan and implement programs properly (MoHUA, 2018).

6.2.1 Study Area

Bangalore is the capital of Karnataka state and one of the world's top ten fastest-growing cities (Sudhira et al., 2007). Between 1949 and 2012, the city's area grew tenfold, with a 632% growth in built-up land use (Ramachandra et al., 2012). According to the global cities study by Oxford Economics (Zhao et al., 2017), the city's yearly economic growth is 8.5%, which is significant in the state's economy. Bengaluru is located between 12° 47' N and 13° 09' N, and 77° 27'E and 77° 47'E, as shown in Figure 6.1, and covers an area of roughly 740 km^2. According to the 2011 census, the city's population was 8.64 million, up 1.5 million from the previous census in 2001. According to prominent scientific assessments, the city has grown rapidly in the last two decades (Sudhira et al., 2003). Bangalore has grown tremendously as a metropolis since 1949. (Ramachandra and Mujumdar, 2009). In five decades, the city's size grew from 69 km^2 to 741 km^2, a tenfold increase in area. The city's current population density is at 12,000 people per square kilometres. The Bangalore Urban District has a population density of 4,378 people per square kilometres, according to the 2011 Census of India. Bangalore's expansion is likely to continue in the following decades with current urbanization trends. In its restructuring report, the city governing authority Bruhat Bengaluru Mahanagara Palike (BBMP) stated that the BBMP jurisdiction would not be sufficient for effective governance until 2025. According to the report,

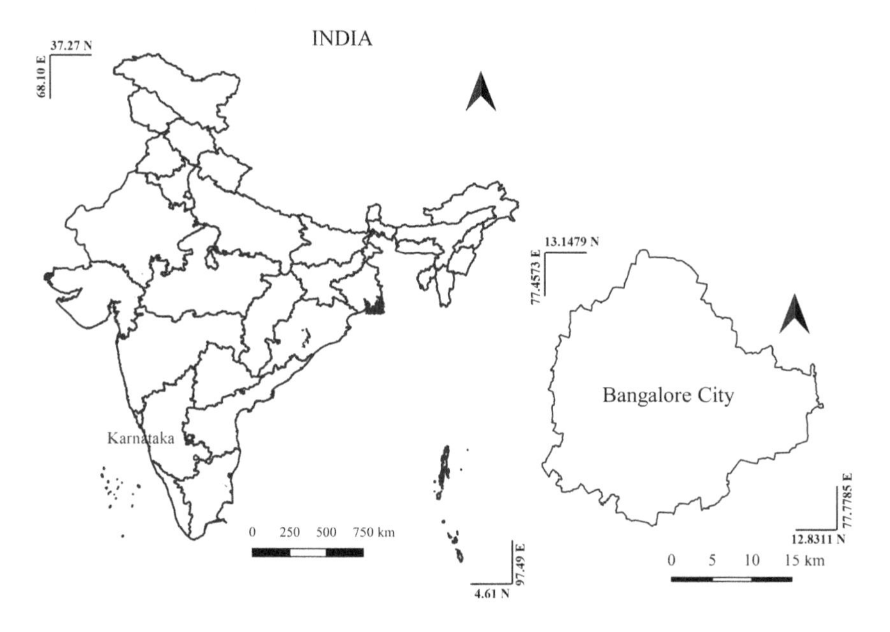

FIGURE 6.1 Study area – Bangalore City, Karnataka state, India.

Bangalore Urban District has a significant urban settlement (Bruhat Bengaluru Mahanagara Palike, 2017).

Bangalore has a pleasant climate due to its geographic location, which includes multiple lakes, natural drainage systems, and ample rainfall. The average annual rainfall is roughly 800 mm, and the city's temperature ranges from 15°C to 35°C for the majority of the year. As a result, Bangalore receives more than 5.5 kWh/m^2/day of Direct Normal Insolation (DNI) on an annual basis (Ramachandra et al., 2011). Bangalore was one of the first few Indian cities to be electrified, with the Kingdom of Mysore electrifying it in 1906. Bangalore is home to about 15% of the state's population, while the city consumes 30% of the state's electricity (Ramachandra et al., 2017). According to the electrical supply firm, the average daily power usage is close to 6 GW, and it is likely to rise substantially in the near future. The BBMP, the city's administrative body, collects property taxes from about 1.3 million individual properties; however, the total number of properties could be as high as 2.5 million. Bangalore is home to 40% of India's IT industry and various other industries such as automobile and textile production (Bala Subrahmanya et al., 2017).

In a paper titled 'Spatial Center Recommendations', a BBMP restructuring expert committee drew out a framework for the Bengaluru Spatial Information Center (Bruhat Bengaluru Mahanagara Palike, 2017). According to the studies, the government spends a significant amount of money to create necessary datasets for various utility agencies. According to the research, building footprints and land-use maps prepared by two agencies for the same area are also inconsistent within multiple spatial data layers. According to the Bangalore Development Authority (BDA), there are approximately 10,00,000 buildings, and according to Bangalore Electricity Supply Company Limited, there are 37,220 buildings (BESCOM). This chapter points out that duplication of data costs, GIS readiness, and the availability of up-to-date spatial information are all key roadblocks to achieving the stated goals.

6.2.2 Datasets

The National Remote Sensing Center (NRSC), a nodal institution of the Indian Space Research Organization (ISRO) for remote sensing and earth observation data, provides satellite images. The TripleSat sensor produces high-resolution satellite imaging with one panchromatic band and four multispectral bands: red, green, blue, and infrared. The panchromatic band has an 80-cm spatial resolution, whereas multispectral bands have a 3.6-m spatial resolution. Additionally, the image contains 10 bits of radiometric data. Surrey Satellite Technology Ltd (SSTL) launched the

TripleSat constellation from Dhawan Space Center in Sriharikota, India, on July 10, 2015. DMC International Imaging Ltd of Guildford, Surrey, UK, owns the spacecraft constellation.

Procurement of remote sensing images, resolution merge or pan-sharpening, development of true colour composite, and labelling of building outlines are all part of the data preparation for extracting building footprints in the study. Building outlines are manually digitized, and then essential building masks are created using rasterization techniques. The image and its matching building mask serve as the deep learning model's input dataset.

6.2.3 Method

Figure 6.2 depicts the approach designed for building extraction work, separated into three parts. The first step is to create and train a CNN-based image segmentation algorithm that divides the image into two classes: building and non-building. Figure 3.2 (Chapter 3) depicts a model architecture that has been trained for Bangalore. Of image tiles, 80% are used as training data during the model-building process, while the remaining 20% are used for monitoring and validation. The trained models are used to predict building pixels in the second stage; however, employed a new set of images that were not exposed to the model during training. Finally, confusion matrix parameters are calculated by comparing expected and real building labels.

6.2.4 Results and Conclusions

The model training is designed for cities with varied built environments. The model offers functionality for fine-tuning hyperparameters to improve forecast accuracy. The images utilized for predictions are not exposed to the model during the training stage to avoid overfitting. The multispectral imagery dataset is required for model training and testing, is created manually as binary building masks. The transfer learning technique is used to train the model iteratively in the GPU system to

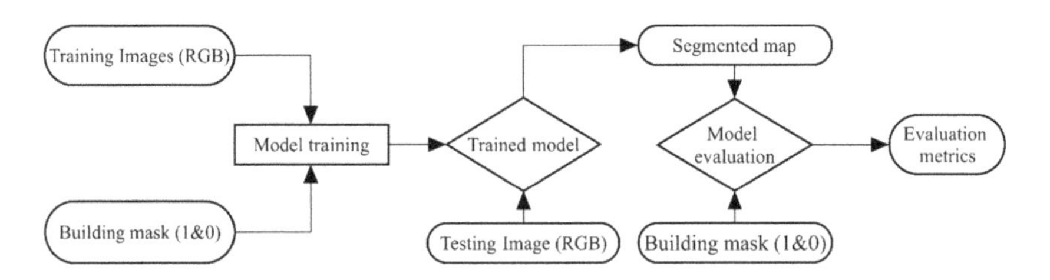

FIGURE 6.2 The method adopted for building extraction.

 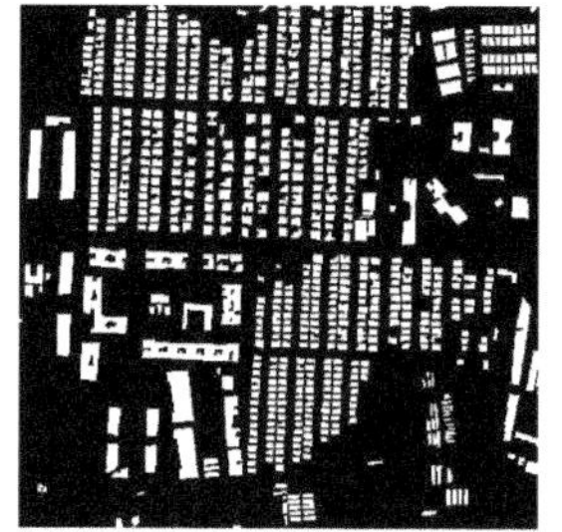 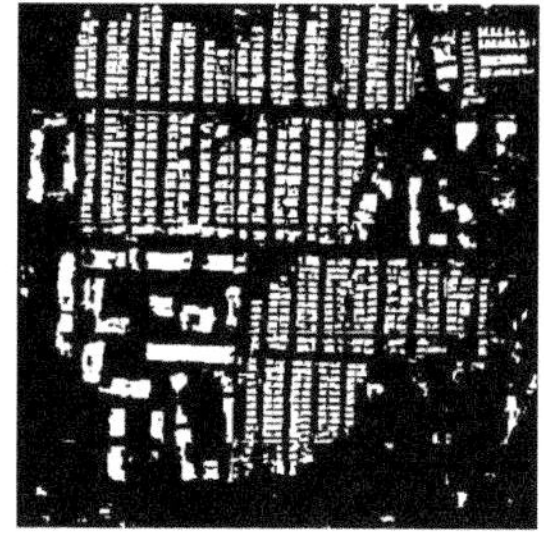

FIGURE 6.3 A sample image, corresponding building mask, and model prediction.

achieve better results. The model's output is the same size, shape, coordinate reference system, and datum as the input image, making subsequent analysis of the results easier, as shown in Figure 6.3. The models are created using the Anaconda open-source software in a Python environment. The results show that the model outputs are better than other contemporary standard architecture at extracting building features quickly. The visual analysis and the metrics of the confusion matrix strongly support this. This approach could aid in creating building footprint databases for a variety of applications. However, because the model gives finer outputs in previously trained scenarios, it is most useful in change detection. The architecture can be enhanced more in the future to make forecasts at the corners of buildings more precise. A tightly compacted construction, such as a house and shed, could need to be segmented individually as an example. Furthermore, the model's performance must be evaluated for more diverse neighbourhoods or large geographic areas. Higher computation capabilities are required to investigate transfer learning strategies in large urban settings (Table 6.1).

TABLE 6.1
The Accuracy Scores of the Segmentation Algorithms and Training Time

Models	Cities	Accuracy	Precision	Recall	F1-Score	Runtime (S)
Proposed U-Net	Test area A	0.96	0.90	0.94	0.92	2811
	Test area B	0.97	0.93	0.86	0.89	5261
	Test area C	0.98	0.94	0.92	0.93	1843
ResNet	Test area A	0.95	0.88	0.91	0.90	2565
	Test area B	0.96	0.85	0.80	0.82	4787
	Test area C	0.97	0.93	0.87	0.90	1690

6.3 CASE STUDY #2: ROOFTOP SOLAR POTENTIAL ESTIMATION

The fast growth of cities necessitates the implementation of re-development plans to meet the government's requirements. The Green Energy Project is one of the few flagship programs that has piqued the interest of stakeholders. By the end of 2022, the Government of India decided to have 40 Giga Watts (GW) of solar rooftop photovoltaic installed (Goel, 2016). Of rooftop solar, 2.33 MW has been physically achieved, according to the Ministry of New and Renewable Energy (MNRE). According to Saha et al. (2016), rooftop solar PV accounts for just around 6% of India's total grid-connected solar energy installed. The Bengaluru urban region, often known as Greater Bangalore, uses 2,000 MW of electricity daily, with peak demand reaching 6,000 MW in the summer.

Several recent investigations have indicated that Karnataka's energy crisis is among the worst in India due to transmission losses. The job of distribution businesses is made more difficult by the presence of many industries and large domestic energy demands. Mahtta et al. (2014) created a solar potential map around the country. According to this report, Karnataka can generate 312 GW of electricity, making it the sixth largest in the country. Ramachandra et al. (2011) mapped solar potential and described these as areas where solar technology might be used to generate electricity throughout the year. Karnataka is one of the states that has implemented net metering, allowing city residents and building owners to benefit from supplying and drawing power from the grid.

Researchers have employed remote sensing data to measure the rooftop solar photovoltaic potential by taking into account factors like roof area and climatic conditions in previous studies (Song et al., 2018). Izquierdo et al. (2008) used remote sensing methods incorporating sampling approaches and digitalization of buildings from Google Earth to assess rooftop solar potential. By digitizing the rooftop areas of residences in selected villages, Ramachandra et al. (2011) calculated rooftop photovoltaic potential. Singh and Banerjee (2015) used a thorough land-use map to make an educated guess regarding the size of the building footprint. For the purpose of estimating roof area, Kapoor et al. (2018) digitized village boundaries and rooftops. These techniques, on the other hand, are exceedingly time-consuming and may result in manual errors during the digitizing process. Vardimon et al. (2011) assessed solar photovoltaics using automated building extraction algorithms from orthophoto. Several machine learning techniques have been utilized to get building polygons over the years. For example, Turker and Koc-San (2015) and Dixon and Candade (2008) employed support vector machines, while Saha and Mandal (2016) and Benarchid and Raissouni (2013) used object-based segmentation. Despite the fact that these machine learning algorithms outperformed hand digitization, they were unable to anticipate proper building shapes for polygonization. This would result in an inaccuracy in the roof area calculation. Convolutional neural networks (CNNs), a deep learning-based technique, is utilized to extract building roofs after considering these factors and analysing several machine learning techniques.

6.3.1 Solar Radiation

Solar radiation is the electromagnetic radiation that originates from the sun. X-rays, ultraviolet, infrared radiation, radio transmissions, microwaves, gamma waves, and visible light are all examples of electromagnetic radiation. The sun emits around 3.8×10^{33} ergs every second, of which the earth captures only a small percentage (1.4 kilowatts/m^2). Solar radiation that strikes the earth's surface can be captured and turned into usable energy. Sun-generated power is considered one of the cleanest forms of energy. The sun's rays strike the earth's surface at angles ranging from 0° to 90° horizontally due to the earth's spherical form. The least scattering occurs when the sun's rays fall completely vertical (i.e., at 90°), and the energy received per unit area is highest. The elliptical orbit of the earth's rotation brings it closer to the sun at some times of the year, allowing it to receive more solar energy. Summer in the southern hemisphere and winter in the northern hemisphere brings the earth closer to the sun. The amount of sunlight falling on the earth's surface is considerably influenced by the 23.5° tilt of the earth's axis of rotation. The Earth's rotation causes the hourly variation in sunshine. Compared to early morning and late afternoon, the sun is perfectly above midday, resulting in least scattering and most energy received per unit area. This is why solar panels produce the most electricity during the midday hours. Because of the tilt angle, the earth sees a unique sun path. As a result, the same panels produce more electricity in the summer.

The method of pointing a solar panel to the sun is known as solar orientation. This orientation is critical because solar photovoltaic panels receive the most solar energy when placed directly at a 90° angle to the sun's rays. Because the sun's location in the sky changes frequently, this is a difficult task. The path the sun traversers, the sunrise in the east and sunset in the west, must be considered. Throughout the year, the sun's route across the sky changes. To determine the solar energy received at a specific location, it is necessary to know the sun's position throughout the year. Between the summer and winter solstice, between 9 a.m. and 3 p.m., a solar window is defined by the sun's position. The solar window of the sun's course is seen in Figure 6.1. This is significant since the sun emits the maximum energy between 9 a.m. and 3 p.m. It is critical that the solar area not be shaded by any objects for the majority of the year in order to receive maximum energy and generate leading solar photovoltaic electricity. The azimuth and solar altitude angles are used to calculate the sun's position in the sky dome. The sun's azimuth angle is when a plumb line drawn from the sun to the horizon intersects a degree from the north line. Solar altitude or solar elevation refers to the angular height above the horizon. Figure 6.1 depicts a representative image.

The total of diffuse solar radiation and direct beam solar radiation is referred to as global solar radiation. Direct beam solar radiation is solar radiation that reaches the earth's surface without being diffused. The technical terminology for sun radiations is diffuse horizontal irradiance (DHI), direct normal irradiance (DNI), and global horizontal irradiance (GHI). The total direct irradiance (after accounting for the solar zenith angle of the Sun z) and diffuse horizontal irradiance is known as global horizontal irradiance (GHI). The GHI expression is given in Equation 6.1. Solar radiation is measured with an instrument called a pyranometer. For solar electric (photovoltaic) systems, radiation data is also expressed in kilowatt-hours per square meter (kWh/m^2). Watts per square meter (W/m^2) is another way to express direct solar energy (Figure 6.4).

$$GHI = DHI + DNI \times \cos(z) \tag{6.1}$$

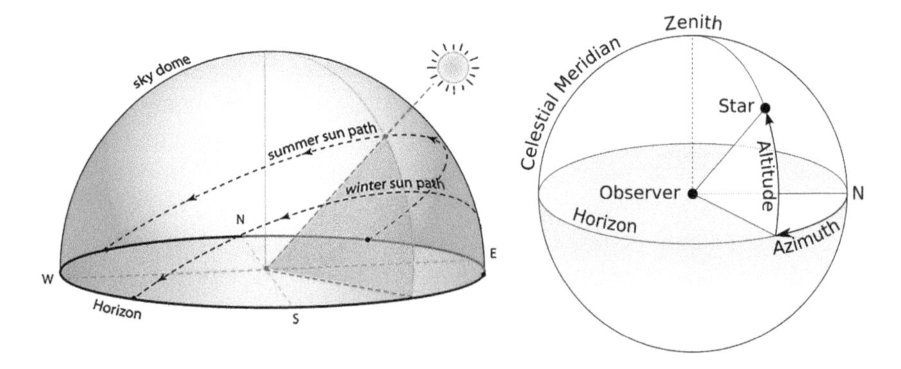

FIGURE 6.4 Solar window and earth-sun geometry. (Image credit: Jeffrey R. S. Brownson.)

Solar potential is the amount of solar energy that can be harnessed by employing photovoltaic systems to generate electricity at a specific location. The solar technological potential is affected by sun angle, air mass, season, day length, temperature, humidity, cloud cover, pollution levels, and other factors in a specific area. The technical solar potential is the amount of power generated in a particular area from photovoltaic panels installed in all appropriate regions. 'Rooftop appropriateness can be estimated using three alternative approaches', according to Margolis et al. (2017): 'constant value methods, human selection, and method based on geographical information systems'. The constant value technique assumes that a percentage of each building's rooftop space is suitable for solar deployment and then multiplies that percentage by the entire roof area of the building. This method estimates the rooftop potential but does not consider the unique characteristics of each building. Using data from aerial imaging or high-resolution satellite photos, manual selection involves evaluating the suitability of each building. The evaluation results are exact, but the procedure is time-consuming and requires a lot of manual effort to apply.

6.3.2 UAV or Drone-Captured Imagery

This case study uses an orthophoto and DSM derived from drone-captured photos for roof area estimation. Figure 6.5 shows the spatial resolution of orthophoto and DSM, which is 20 cm. The image is 1147 × 1147 pixels in size (15 acres). The chosen area is primarily made up of building roofs and is located in a densely populated urban area. Figure 6.6 depicts the four stages of the overall process. In the first stage, images acquired with a quadcopter drone are used to create an orthophoto

FIGURE 6.5 The case study's orthophoto and digital surface model.

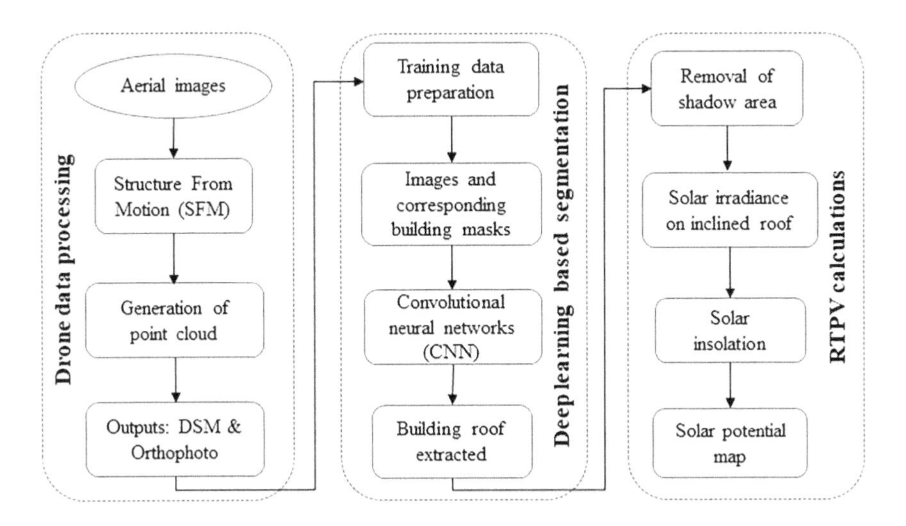

FIGURE 6.6 The energy estimation method using UAV captured imagery.

and DSM. Second, the buildings are retrieved from the orthophoto using a CNN model. The final stage involves removing shadows from the roofs of buildings that have been cast by nearby buildings, trees, and other elevated objects. Solar photovoltaic potential estimations on each building roof are completed in the final stage.

6.3.3 Building Roof Extraction

Building roof areas are calculated using the orthophoto acquired from drone imagery for the study region. The image is segmented into building and non-building classes using a CNN model. A binary construction mask that is manually prepared is used to train the model. Figure 6.7 depicts the vectorization of the buildings predicted by the model utilizing geospatial data abstraction library features.

6.3.4 Shadow Removal

The cityscape was chosen with the goal of having closely spaced buildings of varying heights and trees in the scene. Because of probable shadows from neighbouring buildings and trees, the roof area derived from the model cannot be utilized directly to calculate the solar photovoltaic potential. To begin, the sun's path is projected for the chosen location using the latitude and longitude information to remove the shadows from the rooftops. The sun's path is depicted in Figure 6.8. Then, shadow masks are created for each of the four conditions below, and the roof area that falls under any of the combinations is deleted.

FIGURE 6.7 The prediction and corresponding polygonized roofs overlaid on the imagery.

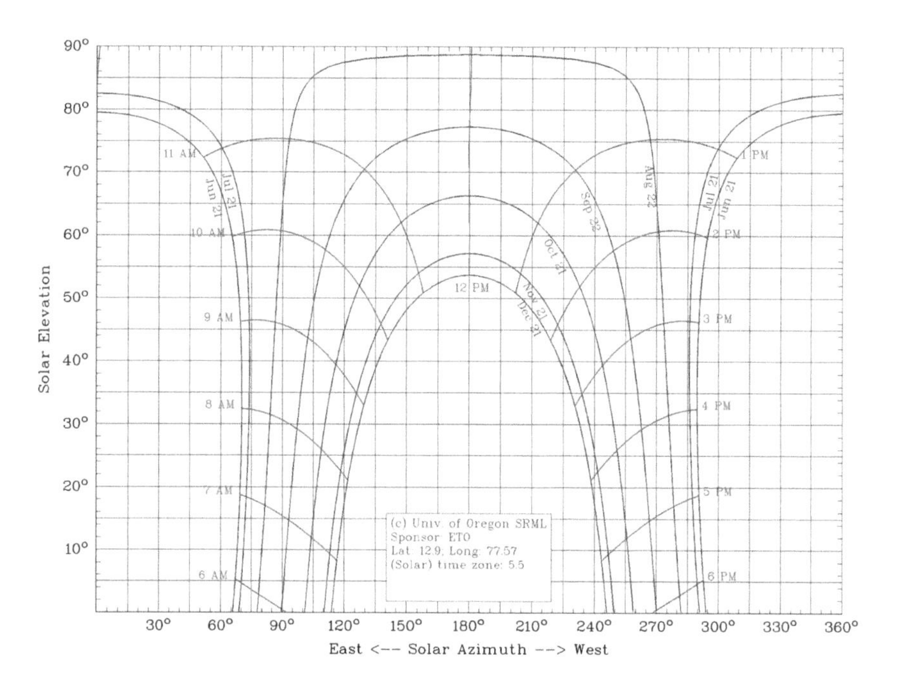

FIGURE 6.8 Sun path chart for the location chosen.

This guarantees that the darkened part of the roof is not taken into account for solar photovoltaic calculations during the solar window (9 a.m. to 3 p.m.). A digital surface model, sun's azimuth angle, and solar elevation angle are used to create shadow masks. Figure 6.9 depicts the shadows obtained for four different extreme circumstances. The shadows are combined, and the shadow area on the roof is erased. After building

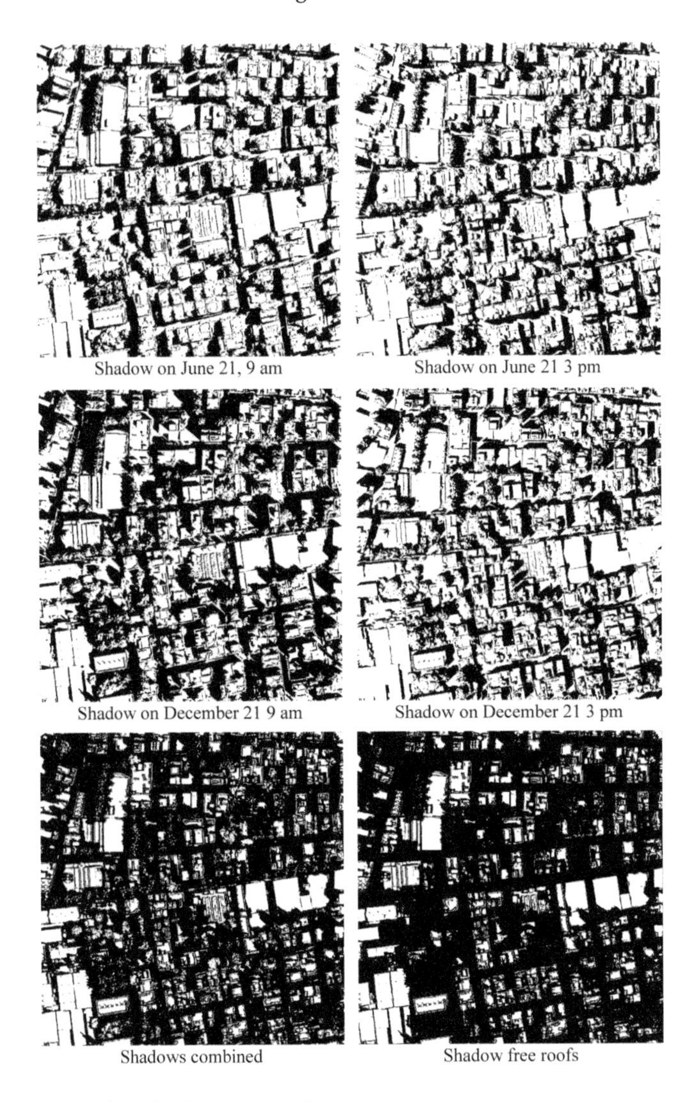

FIGURE 6.9 The shadow removal process.

extraction, the overall area was 62,500 m^2, with a roof surface of 32,585 m^2. Following the removal of the shadow, a total area of 13,056 m^2 is available for solar photovoltaic systems. In other words, solar photovoltaic panels can be installed on up to 40% of the entire roof area (Table 6.2).

6.3.5 Energy Estimations

Table 6.3 shows the parameters that were used to calculate solar insolation values. The sunlight hours and irradiance on the inclined

TABLE 6.2
Solar Azimuth and Elevation Values

Date and Time	Solar Azimuth	Solar Elevation
June 21, 9 a.m.	65	46
June 21, 3 p.m.	295	46
December 21, 9 a.m.	130	33
December 21, 3 p.m.	230	33

TABLE 6.3
Parameters Considered to Calculate Irradiance on an Inclined Surface

Parameters	Values
Coordinates	77.573 E, 12.897 N
Elevation	910 m
Azimuth	65°–230°
Inclination	20°
Albedo	0.15
Panel efficiency	0.15

surface for each month are determined for the location in question. In addition, as illustrated in Figure 7.9, the energy potential of each building's roof is mapped. Table 6.4 shows the sun irradiation value on an inclined surface and energy potential estimates (Figure 6.10).

The following information can be found on the solar potential map created for the case study. Six buildings have a power production of more than 31 kWh, nine buildings have a power generation of 13–31 kWh, 29 buildings have a power production of 5–13 kWh, and 144 buildings have a power production of 1–5 kWh. Out of a total of 302 buildings in the study region, the remaining 144 use less than 1 kWh. The calculations above are for the month of January. The study area's average energy potential is 0.82 MW during the full year. This is computed by multiplying monthly generation by sunshine hours, energy on an inclined surface, and total roof area accessible for solar installation. According to estimates, the energy output is predicted to be highest in December and January and lowest in June.

TABLE 6.4
Rooftop Solar Photovoltaic Energy Potential

Month	Irradiance Inclined Surface (kWh/m^2/day)	Sunshine (Hours)	Potential Energy Output (kWh/day)	Energy Potential (MW)
January	7.94	9.4	9,952	1.06
February	7.93	10.4	9,944	0.96
March	7.13	10.1	8,941	0.89
April	6.47	10.1	8,110	0.80
May	5.58	9.5	6,999	0.74
June	3.94	8.3	4,936	0.59
July	4.00	7.5	5,017	0.67
August	4.19	7.2	5,259	0.73
September	4.47	8.4	5,602	0.67
October	5.65	8.5	7,080	0.83
November	5.93	8.6	7,441	0.87
December	7.45	8.8	9,346	1.06
Annual average				0.82

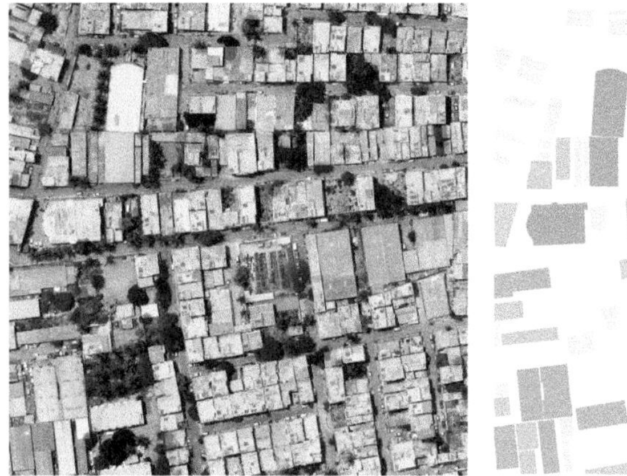

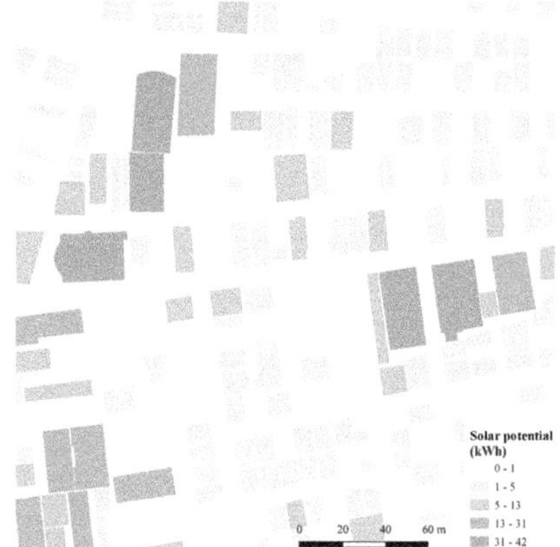

FIGURE 6.10 Roofs and solar potential map for the month of January.

The major goal of this case study is to show how to extract building roof information from satellite images and estimate solar potential using geospatial technology and machine learning approaches. The case studies demonstrate the value of deep learning approaches for estimating the roof area of a structure. Manual digitization, field-based data collecting, and sampling approaches are all slower than the proposed solution.

The method for determining a city's or region's energy potential is scientific and data-driven. However, there is room for improvement in making the estimations more robust. The proposed methods do not properly capture rooftops made of materials such as brick and other coloured materials. The sun-earth geometry of the digital surface model aids in precise roof area estimations. Despite the fact that the estimates in the second case study are for the worst-case scenario, actual production is predicted to be slightly higher than the estimated figure.

6.4 CASE STUDY #3: ASSESSMENT OF URBAN BUILT-UP VOLUME

Geospatial technology has proven to be an effective tool for accurately measuring and monitoring spatial phenomena. The spike in data availability, constantly improving the spatial resolution of images, development of automated methods, and significant increases in processing capabilities can all be attributable to the increased interest of the scientific community in using remote sensing data. Because of the dense, complex, and varied nature of urban form, mapping urban areas is a complicated process. Understanding and visualizing changes in land use is critical in urban studies because it provides policy instruments and is a critical parameter in sustainable urban development. Though these studies provide a pattern of growth, they cannot quantify the volume of growth since vertical growth is frequently overlooked and crucial for resource and policy planning. Economic superiority, high-density zones in terms of population, and higher resource consumption can all be measured using vertical growth. As a result, height information must be integrated for increased clarity. Analysis can be carried out using available remotely sensed datasets such as elevation models. The intricacy of an urban zone is not captured by publicly available medium resolution data such as ASTER/SRTM/Cartosat-1/ALOS PRISM (Misra et al., 2018).

This involves using high-resolution data, such as Cartosat 1 stereo images, which provide DSMs with an accuracy of up to 2 pixels in planimetry and 4 pixels vertically when accurate GCPs are included. Traditional classifiers are not capable of recovering building footprints from high resolution satellite images in the event of complex urban scenarios, hence with higher resolution comes higher classification accuracy. When it comes to recognizing small features such as buildings, deep learning has proven to be effective and accurate. Deep learning algorithms require a large number of labelled datasets to be trained. To extract urban volume information, this case study combines the use of a deep learning model with high-resolution data.

FIGURE 6.11 LISS-IV (left) and Cartosat-1 (right) stereo images were used in the study.

6.4.1 Study Area and Datasets

Bangalore, India's Silicon Valley, was chosen for this case study because it has experienced fast population growth and land-use change over the last two decades. All urban forms are present in a complex and dynamic manner. Figure 6.1 depicts the study region (741 km^2). The study uses Cartosat-1 stereo pictures to create a DSM and calculate the height component. Built-up area is extracted using LISS-IV MSS data with a resolution of 5.8 m. To execute automatic triangulation, accurate ground control points collected using differential GPS are employed during geo rectification, accuracy assessment, and tie point production (Figure 6.11).

6.4.2 Method

As indicated in Figure 6.12, the overall process consists of three stages. DSM is created from stereo images and ground control points in the first stage using a photogrammetric technique. The built-up area is calculated

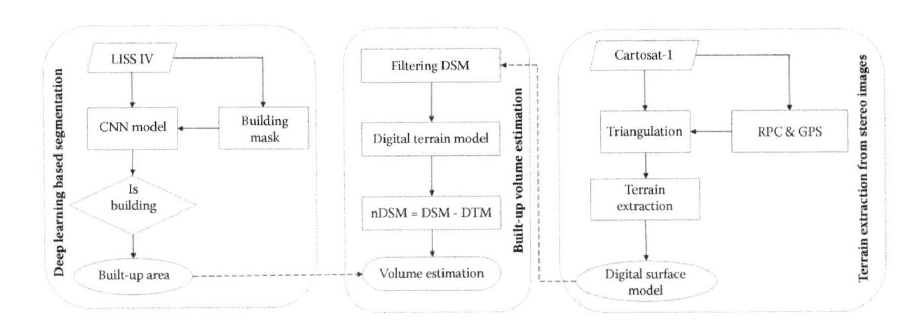

FIGURE 6.12 Flowchart showing the method adopted.

using deep learning techniques and a high-resolution satellite image in the second stage. Built-up volume calculations are made in the last stage. The method utilized for each stage is described in detail in the following sections.

6.4.3 DSM Generation

The Rational Polynomial Coefficient (RPC) sensor model is provided with Cartosat-1 stereo satellite images. The RPC sensor model is created using the satellite's altitude and orbit data. The RPC sensor model is less accurate than the image's spatial resolution, which is 2.5 m. As a result, the model must be corrected using precise ground control points, which can be done via affine transformation (Lehner et al., 2006), as shown below:

$$row = a_0 + a_1 RPC_{row} + a_2 RPC_{col} \tag{6.2}$$

$$col = b_0 + b_1 RPC_{row} + b_2 RPC_{col} \tag{6.3}$$

where RPC_{row} and RPC_{col} are rational polynomial functions with stereo images in this case.

Block triangulation is used to establish a mathematical relationship between stereo images in the model and the ground. The approach known as bundle block adjustment is used to process all of the images in the model simultaneously. The procedure is carried out with the help of RPC bundle block adjustment, which is based on the equations below (d'Angelo, 2013):

$$\Delta r_{ji} = r_j(\varphi_i, \Theta_i, h_i) - r_{ji*} \tag{6.4}$$

$$\Delta c_{ji} = c_j(\varphi_i, \Theta_i, h_i) - c_{ji*} \tag{6.5}$$

where φ_i, Θ_i, and h_i are object coordinates of point i; r_j and c_j are projection functions for row and column from object coordinates to image coordinate. r_{ji*} and c_{ji*} are the measured image coordinates in image j.

Satellites with affine transformation and precise ground control points can produce high-quality results. If ground coordinates are considered as φ_{i*}, Θ_{i*}, h_{i*}, additional observations are added as follows:

$$\Delta\varphi_i = \varphi_i - \varphi_{i*} \tag{6.6}$$

$$\Delta\Theta_i = \Theta_i - \Theta_{i*} \tag{6.7}$$

$$\Delta h_i = h_i - h_{i*} \tag{6.8}$$

A tie point on an overlapping image region is a recognizable location between stereo images where thecoordinates are unknown. Tie points are formed based on the geometrical and radiometric properties of the stereo images. Least square matching is used to confirm the quality of the tie points generated. The tie points are kept within 0.3 pixels of positional accuracy. To create a dense disparity map, a stereo pair of images and tie points are matched using image matching techniques (d'Angelo and Reinartz, 2011). The DSM was created by combining the disparity map and projection into the required coordinate system. Figure 6.13 depicts the DSM that was obtained. DSM's resolution is adjusted to 5 m based on observations from previous studies (Deilami and Hashim, 2011). To differentiate above-ground items from the bare earth surface, slope-based filtering is performed to the DSM produced. The bare ground surface is interpolated to create a digital terrain model, an elevation model that only

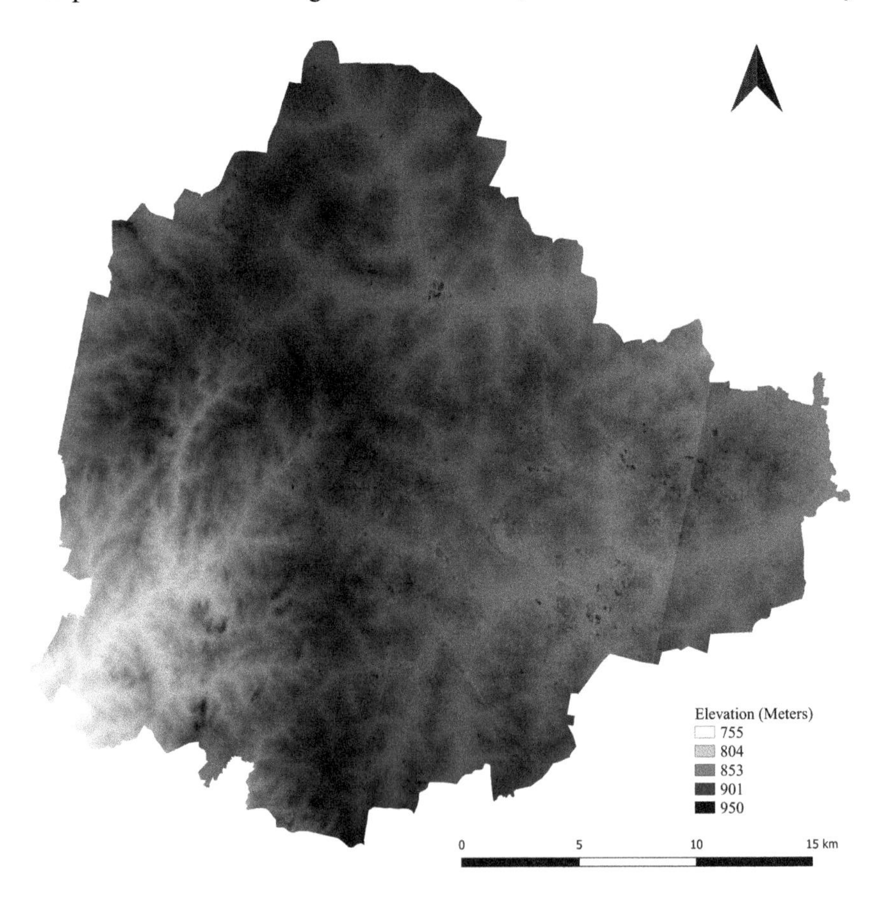

FIGURE 6.13 For the entire city, the digital surface model is created.

comprises terrain surface (DTM). The height map, or nDSM (normalized DSM), is created by subtracting the DTM from the DSM and contains the height of the above-ground objects.

6.4.4 Built-Up Area Extraction

The built-up area from the study is extracted using LISS-IV satellite images. Initially, building masks are chosen at random from diverse parts to account for the variability of land-use patterns. Building masks have a '1' value for buildings and a '0' value for the background. With the use of image bands and a corresponding building mask, a CNN model is trained. For training, the CNN model uses building masks and three bands of imagery as input data. A robust deep CNN model termed U-Net is constructed to divide the image into two classes. Figure 6.14 depicts the subset of imagery, related masks, and resulting predictions. The built-up area that has been extracted is utilized to calculate the built-up volume.

FIGURE 6.14 For the entire city, the model estimated the built-up area.

6.4.5 Built-Up Volume Estimation

The built-up volume is calculated using nDSM and built-up area maps. Before doing the volume calculations, a resampling approach is utilized to make the spatial resolution of the nDSM and built-up area map uniform. The formula for calculating the built-up volume can be found below.

$$nDSMx = UBVx = BAx$$

UBV, BA, and nDSM refer to the pixel's urban built-up volume (m^3/pixel), built-up area (25 m^2), and height of above-ground objects (m). Pixel x is a representative member of the built-up class in this case. To show the built-up volume of the city, the built-up volume generated at 5 m^2 is mapped into a vector grid with a grid size of 1 km^2. As indicated in Figure 6.15, the built-up volume is divided into five categories.

FIGURE 6.15 Built-up volume map for the city of Bangalore.

6.4.6 Inference and Conclusions

The results can be analysed in three stages, the first two of which entail the dataset preparation. Three sceneries of Cartosat-1 stereo pairs are employed in the initial stage, associated with 25 ground control points. The DSMs for the three pairs are combined and clipped for the study area. The built-up area map may be generated by using the CNN model to segment the LISS-IV satellite image. Building masks are used to train the CNN model. Open Street Map data is used to create the building masks needed for training and testing the model. The model's prediction is tested against 30% of the dataset's area, achieving an overall accuracy of 89%. Because the model's forecast isn't in binary form, a threshold value is used to convert it into a binary map with built-up and non-built-up areas.

Figure 6.15 shows that the city's eastern areas have a higher built-up volume than the central part, indicating that high-rise buildings are present. Due to the scarce presence of buildings and the indication of spread, the city's outskirts are classified as 'extremely low' built-up volume. 363 grids in the study area have a built-up volume of less than 20,000 m^3/pixel; 248 grids have a built-up volume of between 20,000 and 40,000 m^3/pixel; 145 grids have a built-up volume of between 40,000 and 60,000 m^3/pixel; 20 grids have a built-up volume of between 60,000 and 80,000 m^3/pixel; and 5 grids have a built-up volume of more than 80,000 m^3/pixel This is the city's indicative metric, which can be used to examine the city's changing landscape.

With the help of built-up volume, the growth pattern of cities or urban forms can be effectively comprehended. The urban built-up volume is an important characteristic that can bring a lot of value to the phenomenon of horizontal urban growth. The pattern of urban expansion has shifted in recent years from horizontal growth to a blend of horizontal and vertical growth. This case study demonstrates a step-by-step strategy to studying urban growth patterns. The prepared output can be utilized as a proxy measure for describing economic growth, population mobility patterns, societal typology, environmental indicators, etc. According to the current case study, the city's eastern half has a higher built-up volume than the central city. The method can also aid in identifying potential future development and economic activity regions. Further research could involve a look at the city's built volume over time. Models can also be designed to estimate future growth localities based on the built-up volume and other urban growth factors.

REFERENCES

Aayog, NITI. (2017). Sustainable development goals (SDGs), targets, CSS, interventions, nodal and other ministries. *Development Monitoring and Evaluation Office. Government of India. New Delhi.*

Ahluwalia Committee & Isher, J. (2011). Report on Indian Urban Infrastructure and Services. Delhi: ICRIER.

Bala Subrahmanya, M. H. (2017). How did Bangalore emerge as a global hub of tech start-ups in India? Entrepreneurial ecosystem—evolution, structure and role. *Journal of Developmental Entrepreneurship*, 22(01), 1750006.

Benarchid, O., Raissouni, N., El Adib, S., Abbous, A., Azyat, A., Achhab, N. B., & Chahboun, A. (2013). Building extraction using object-based classification and shadow information in very high-resolution multispectral images, a case study: Tetuan, Morocco. *Canadian Journal on Image Processing and Computer Vision*, 4(1), 1–8.

Bharath, H. A., Chandan, M. C., Vinay, S., & Ramachandra, T. V. (2017). Intra and inter spatio-temporal patterns of urbanisation in Indian megacities. *International Journal of Imaging and Robotics*, 17(2), 28–39. http://wgbis.ces.iisc.ac.in/energy/water/paper/Intra-and-Inter-Spatio-Temporal-Patterns/sec9.html

Bruhat Bengaluru Mahanagara Palike. (2017). *Bengaluru: Way Forward. BBMP Restructuring.* https://opencity.in/documents/bengaluru-way-forward-bbmp-restructuring-2015

Chandan, M. C., Nimish, G., & Bharath, H. A. (2020). Analysing spatial patterns and trend of future urban expansion using SLEUTH. *Spatial Information Research*, 28(1), 11–23. doi:10.1007/s41324-019-00262-4

d'Angelo, P. (2013, November). Automatic orientation of large multitemporal satellite image blocks. In *Proceedings of International Symposium on Satellite Mapping Technology and Application 2013*, pp. 1–6.

d'Angelo, P., & Reinartz, P. (2011). Semiglobal matching results on the ISPRS stereo matching benchmark. In *Int. Arch. Photogramm. Remote Sens. Spatial Inf. Sci.*, 79–84.

Dash, J., Steinle, E., Singh, R. P., & Bähr, H. P. (2004). Automatic building extraction from laser scanning data: an input tool for disaster management. *Advances in Space Research*, 33(3), 317–322. doi:10.1016/S0273-1177(03)00482-4

Deilami, K., & Hashim, M. (2011). Very high-resolution optical satellites for DEM generation: A review. *European Journal of Scientific Research*, 49(4), 542–554.

Dixon, B., & Candade, N. (2008). Multispectral landuse classification using neural networks and support vector machines: One or the other, or both? *International Journal of Remote Sensing*, 29(4), 1185–1206. doi:10.1080/01431160701294661

Goel, M. (2016). Solar rooftop in India: Policies, challenges and outlook. *Green Energy & Environment*, 1(2), 129–137. doi:10.1016/j.gee.2016.08.003

Harini,B. (2017). *Illegal Constructions: An Unending Battle for the Government and Home Buyer.* https://www.proptiger.com/guide

Isher, J. A. (2011). *The High-Powered Expert Committee (HPEC) for Estimating the Investment Requirements for Urban Infrastructure Services.*

Izquierdo, S., Rodrigues, M., & Fueyo, N. (2008). A method for estimating the geographical distribution of the available roof surface area for large-scale photovoltaic energy-potential evaluations. *Solar Energy*, 82, 929–939. 10.1016/j.solener.2008.03.007.

Kapoor, M., & Garg, R. D. (2018). Cloud computing for energy requirement and solar potential assessment. *Spatial Information Research*, 26(4), 369–379. doi:10.1007/s41324-018-0181-3

Lehner, M., Müller, R., & Reinartz, P. (2006, September). Stereo evaluation of Cartosat-1 data on test site 5-first DLR results. *In 2006 Proceedings of Symposium Geospatial Databases for Sustainable Development.*

Mahtta, R., Joshi, P. K., & Jindal, A. K. (2014). Solar power potential mapping in India using remote sensing inputs and environmental parameters. *Renewable Energy*, 71, 255–262. doi:10.1016/j.renene.2014.05.037

Margolis, R., Gagnon, P., Melius, J., Phillips, C., & Elmore, R. (2017). Using GIS-based methods and Lidar data to estimate rooftop solar technical potential in US cities. *Environmental Research Letters*, 12(7), 074013.

Misra, P., Avtar, R., & Takeuchi, W. (2018). Comparison of digital building height models extracted from AW3D, TanDEM-X, ASTER, and SRTM digital surface models over Yangon City. *Remote Sensing*, *10*(12), 2008

MNRE. (2014). National solar mission. *Grid Connected Solar Rooftop Program in India.*

MoHUA. (2018). Ministry of Housing and Urban Affairs — Mandate. New Delhi.

MoHUA. (2014). Ministry of Housing and Urban Affairs. Available at: https://mohua.gov.in/upload/. Last accessed 30 AUGUST 2022.

MOWR. (2013). Master plan for artificial recharge to ground water in India. *Central, New Delhi.*

Ramachandra, T. V., & Mujumdar, P. P. (2009). Urban floods: Case study of Bangalore. *Disaster Dev*, 3(2), 1–98.

Ramachandra, T. V., & Sudhira, H. S. (2011). Influence of planning and governance on the level of urban services. *IUP Journal of Governance and Public Policy*, 6(1), 24–50.

Ramachandra, T. V., Aithal, B. H., & Sanna, D. D. (2012). Insights to urban dynamics through landscape spatial pattern analysis. *International Journal of Applied Earth Observation and Geoinformation*, 18, 329–343. doi 10.1016/j.jag.2012.03.005

Ramachandra, T. V., Bajpai, V., Kulkarni, G., Aithal, B. H., & Han, S. S. (2017). Economic disparity and CO_2 emissions: The domestic energy sector in Greater Bangalore, India. *Renewable and Sustainable Energy Reviews*, 67, 1331–1344. doi:10.1016/j.rser.2016.09.038

Ramachandra, T. V., Jain, R., & Krishnadas, G. (2011). Hotspots of solar potential in India. *Renewable and Sustainable Energy Reviews*, 15(6), 3178–3186. doi:10.1016/j.rser.2011.04.007

Ranagalage, M., Estoque, R. C., Handayani, H. H., Zhang, X., Morimoto, T., Tadono, T., & Murayama, Y. (2018). Relation between urban volume and land surface temperature: A comparative study of planned and traditional cities in Japan. *Sustainability*, 10(7), 2366.

Saha, K., & Mandal, N. R. (2016). Estimating solar PV potential for sustainable energy planning in tier-II cities of India: Case of Bhopal City. *Current Urban Studies*, 4(03), 356. doi:10.4236/cus.2016.43024

Shetty, A., Umesh, P., & Shetty, A. (2021). An exploratory analysis of urbanization effects on climatic variables: A study using Google Earth Engine. *Modeling Earth Systems and Environment*, 1–16. doi:10.1007/s40808-021-01157-w

Singh, R., & Banerjee, R. (2015). Estimation of rooftop solar photovoltaic potential of a city. *Solar Energy*, 115, 589–602. doi:10.1016/j.solener.2015.03.016

Song, X., Huang, Y., Zhao, C., Liu, Y., Lu, Y., Chang, Y., & Yang, J. (2018). An approach for estimating solar photovoltaic potential based on rooftop retrieval from remote sensing images. *Energies*, 11(11), 3172. doi:10.3390/en11113172

Sudhira, H. S., Ramachandra, T. V., & Subrahmanya, M. B. (2007). Bangalore. *Cities*, 24(5), 379–390. doi:10.1016/j.cities.2007.04.003

Sudhira, H. S., Ramachandra, T. V., Raj, K. S., & Jagadish, K. S. (2003). Urban growth analysis using spatial and temporal data. *Journal of the Indian Society of Remote Sensing*, 31(4), 299–311. doi:10.1007/BF03007350

Turker, M., & Koc-San, D. (2015). Building extraction from high-resolution optical spaceborne images using the integration of support vector machine (SVM) classification, Hough transformation and perceptual grouping. *International Journal of Applied Earth Observation and Geoinformation*, 34, 58–69. doi:10.1016/j.jag.2014.06.016

Vardimon, R. (2011). Assessment of the potential for distributed photovoltaic electricity production in Israel. *Renewable Energy*, 36(2), 591–594. doi:10.1016/j.renene.2010.07.030

Zhao, S. X., Guo, N. S., Li, C. L. K., & Smith, C. (2017). Megacities, the world's largest cities unleashed: Major trends and dynamics in contemporary global urban development. *World Development*, 98, 257–289. doi:10.1016/j.worlddev.2017.04.038

Index